THE
HUMAN
BODY
ATLAS

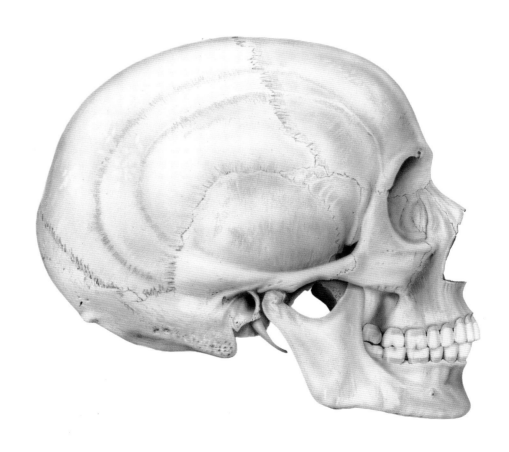

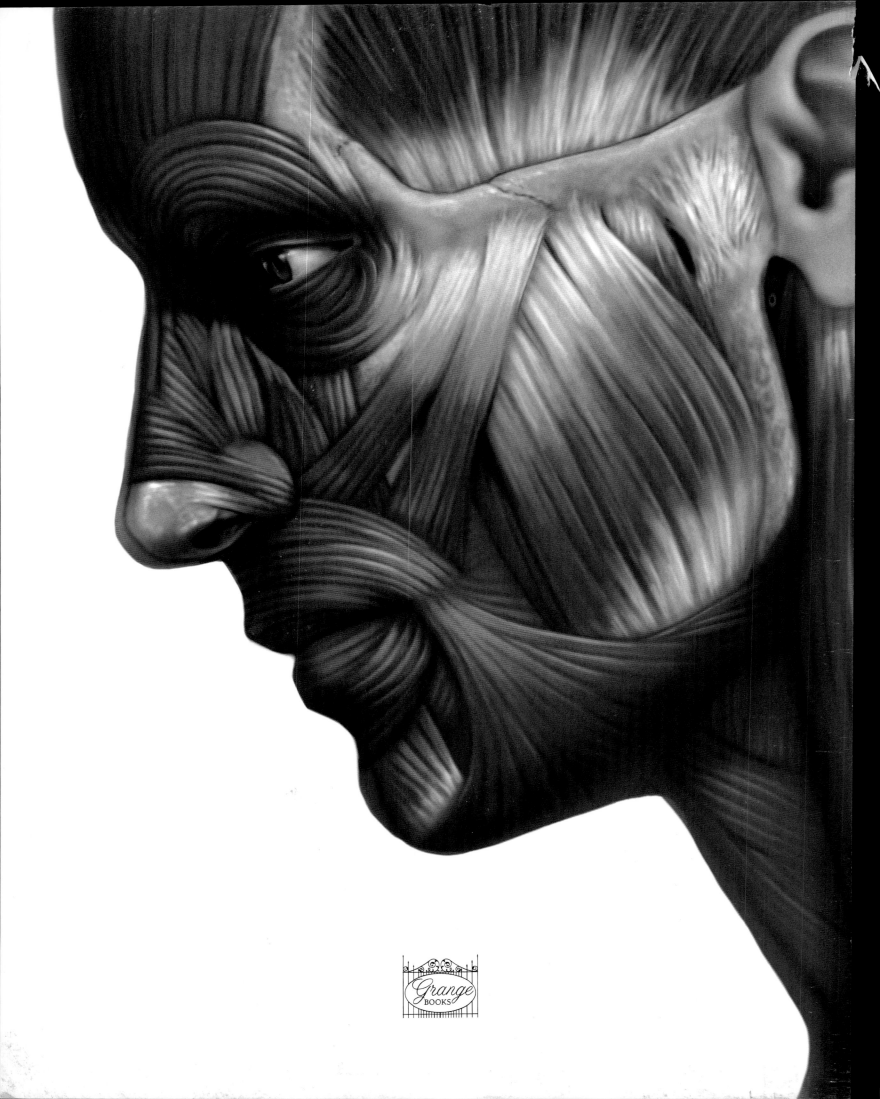

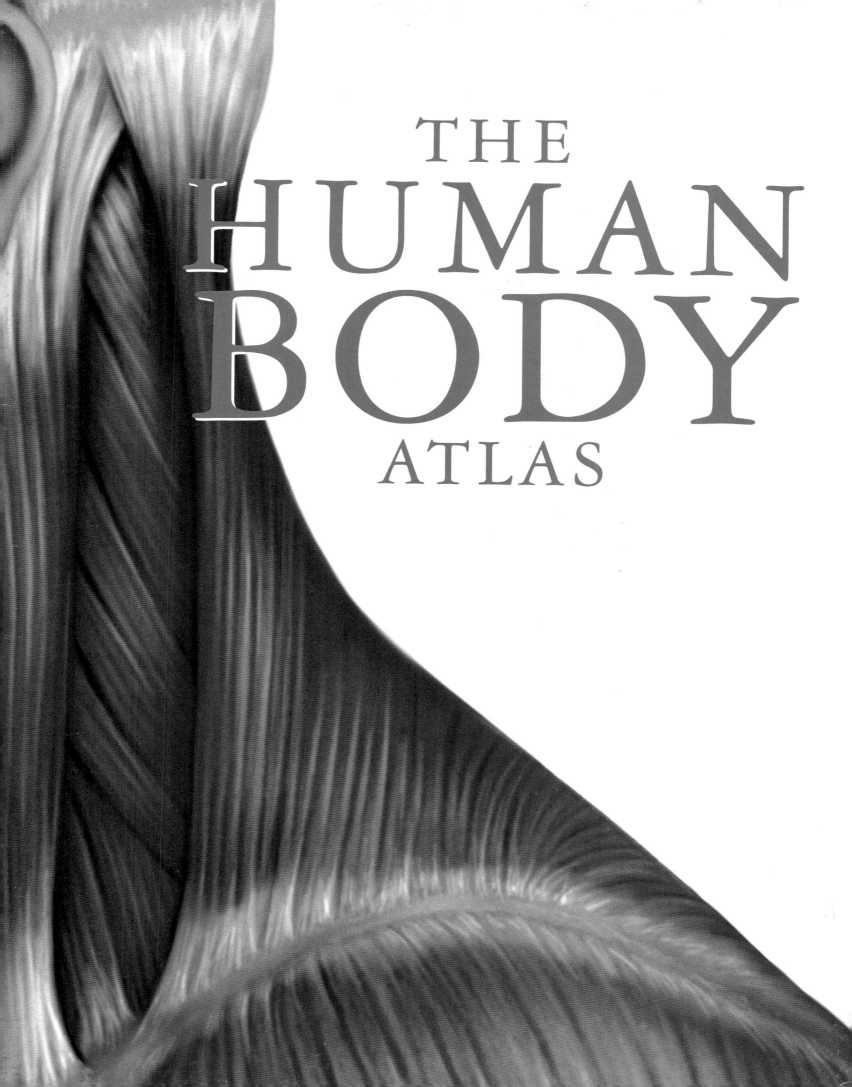

THE
HUMAN
BODY
ATLAS

Publisher	Gordon Cheers
Associate publisher	Margaret Olds
Managing editor	Janet Parker
Chief consultants	Kurt H. Albertine, PhD
	The Honorable Emeritus Professor Peter Baume, AO, MD, BS, HonDLitt, FRACP, FRACGP, FAFPHM
	Dr R. William Currie, BSA, MSc, PhD
	Laurence Garey, MA, DPhil, BM, BCh
	Gareth Jones, BSc (Hons), MB, BS, DSc, CBiol, FIBiol
	David Tracey, BSc, PhD
Contributors	Robin Arnold, MSc
	Ken Ashwell, BMedSc, MB, BS, PhD
	Deborah Bryce, BSc, MScQual, MChiro, GrCertHEd
	Carol Fallows, BA
	Martin Fallows
	John Frith, MB, BS, BSc(Med), GradDipEd, MCH, RFD
	John Gallo, MB, BS (Hons), FRACP, FRCPA
	Brian Gaynor, MB, BS, FRACP
	Rakesh Kumar, MB, BS, PhD
	Peter Lavelle, MB, BS
	Lesley Lopes, BA Communica
	Karen McGhee, BSc
	Michael Roberts, MB, BS, LLB
	Emeritus Professor Frederick PhD, DCP, DipRMS
	Elizabeth Tancred, BSc, PhD
	Dzung Vu, MD, MB, BS, DipA
	Phil Waite, BSc (Hons), MBCh
Chief illustration consultant	Dzung Vu, MD, MB, BS, DipA
Illustration consultants	John Frith, MB, BS, BSc(Med)
	David Jackson, MB, BS, BSc(Med)
Illustrators	David Carroll
	Peter Child
	Deborah Clarke
	Geoff Cook
	Marcus Cremonese
	Beth Croce
	Wendy de Paauw
	Levant Efe
	Hans De Haas
	Mike Golding
	Jeff Lang
	Alex Lavroff
	Ulrich Lehmann
	Ruth Lindsay
	Richard McKenna
	Annabel Milne
	Tony Pyrzakowski
	Oliver Rennert
	Caroline Rodrigues
	Otto Schmidinger
	Bob Seal
	Vicky Short
	Graeme Tavendale
	Jonathan Tidball
	Paul Tresnan
	Valentin Varetsa
	Glen Vause
	Spike Wademan
	Trevor Weekes
	Paul Williams
	David Wood

Text editors	Alan Edwards
	Denise Imwold
	Janet Parker
Art director	Stan Lamond
Cover design	Andrew Davies
	Stan Lamond
Book design concept	Andrew Davies
Page layout	Paula Kelly
	Dee Rogers
Typesetting	Dee Rogers
Index	Glenda Browne
International rights	Rosemary Barry
Publishing assistant	Erin King
Production	Bernard Roberts

Published by Grange Books
an imprint of Grange Books Plc
The Grange
Kingsnorth Industrial Estate
Hoo, nr Rochester
Kent ME3 9ND
www.Grangebooks.co.uk

ISBN 1 84013 510 7

Produced by Global Book Publishing Pty Ltd
1/181 High Street, Willoughby, NSW, 2068, Australia
Ph (+61 2) 9967 3100 Fax (+61 2) 9967 5891
Email globalpub@ozemail.com.au

First published in 2002

Illustrations from the Global Illustration Archives
© Global Book Publishing Pty Ltd 2002
Text © Global Book Publishing Pty Ltd 2002

Printed in China by Midas Printing (Asia) Ltd
Colour separation Pica Digital Pte Ltd, Singapore

While every care has been taken in presenting this
material, the medical information is not intended to replace
professional medical advice; it should not be used as a guide for
self-treatment or self-diagnosis. Neither authors nor the publisher
may be held responsible for any type of damage or harm caused
by the use or misuse of information in this book.

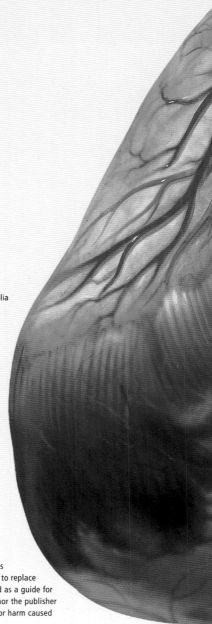

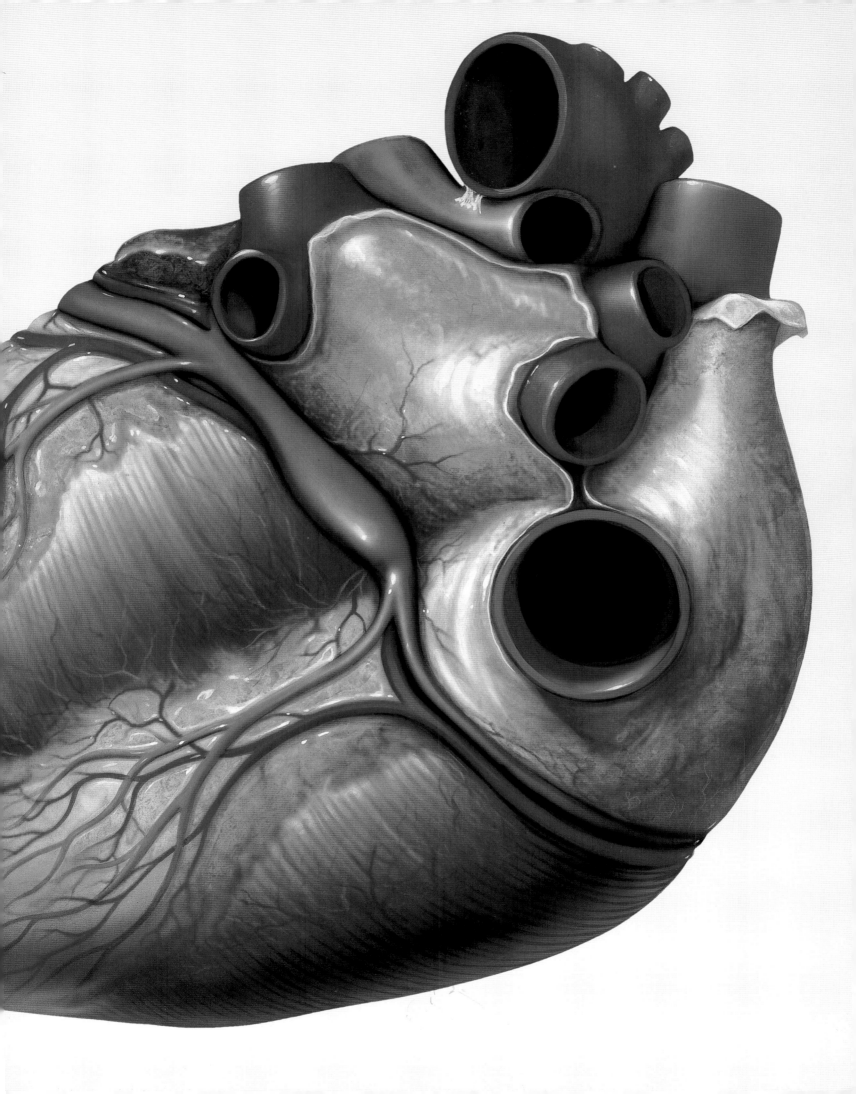

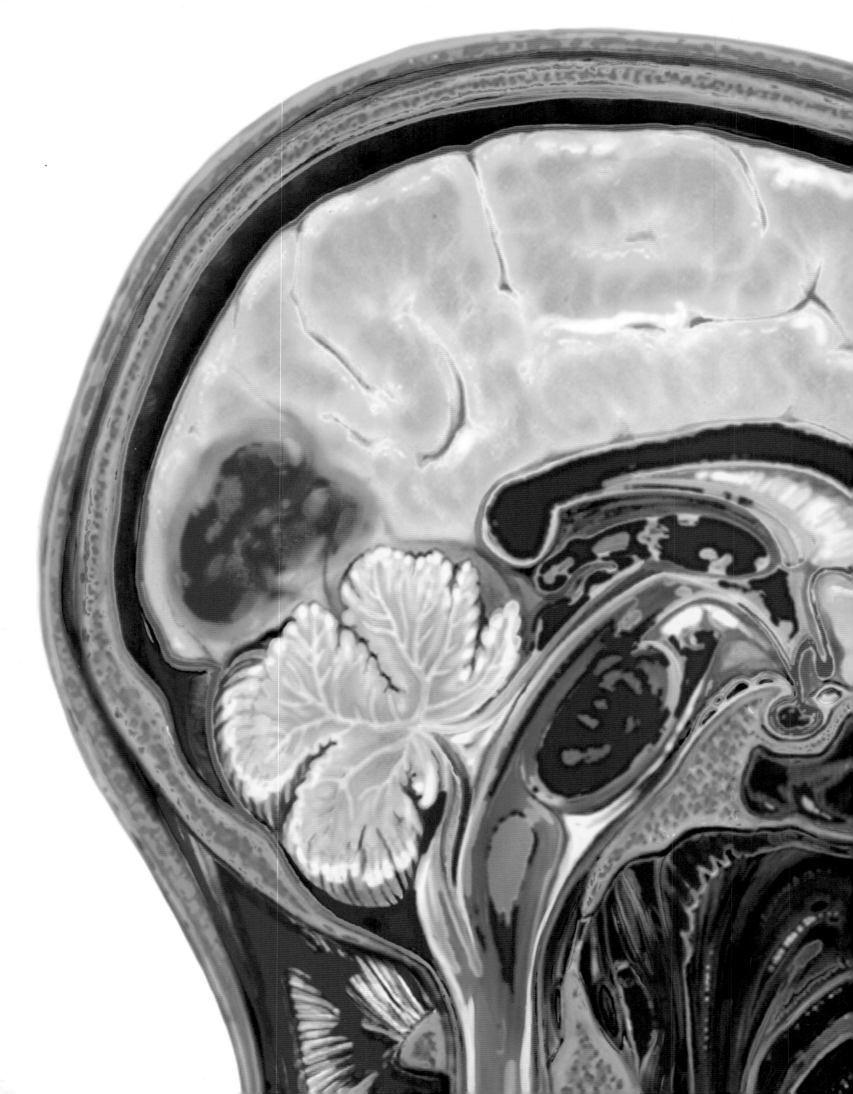

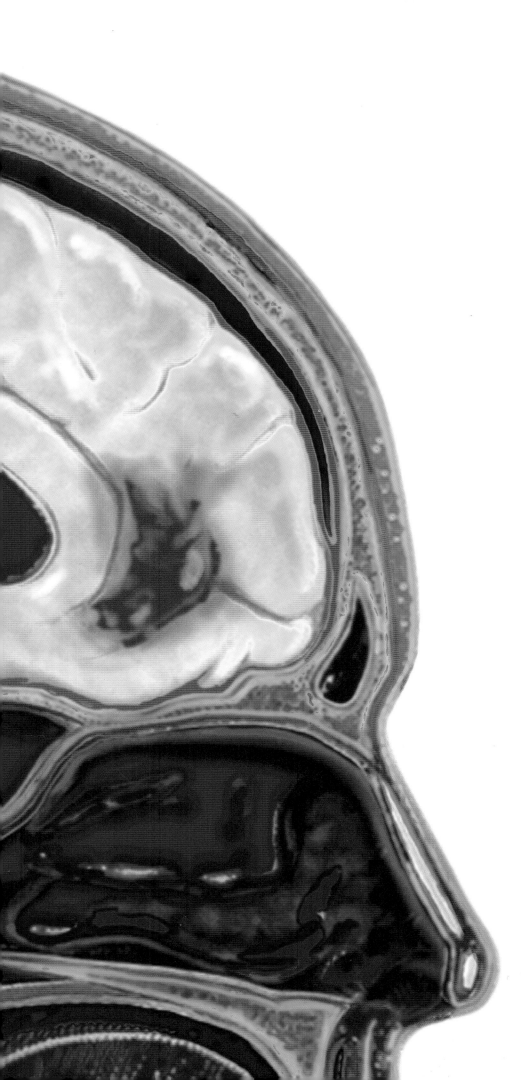

Contents

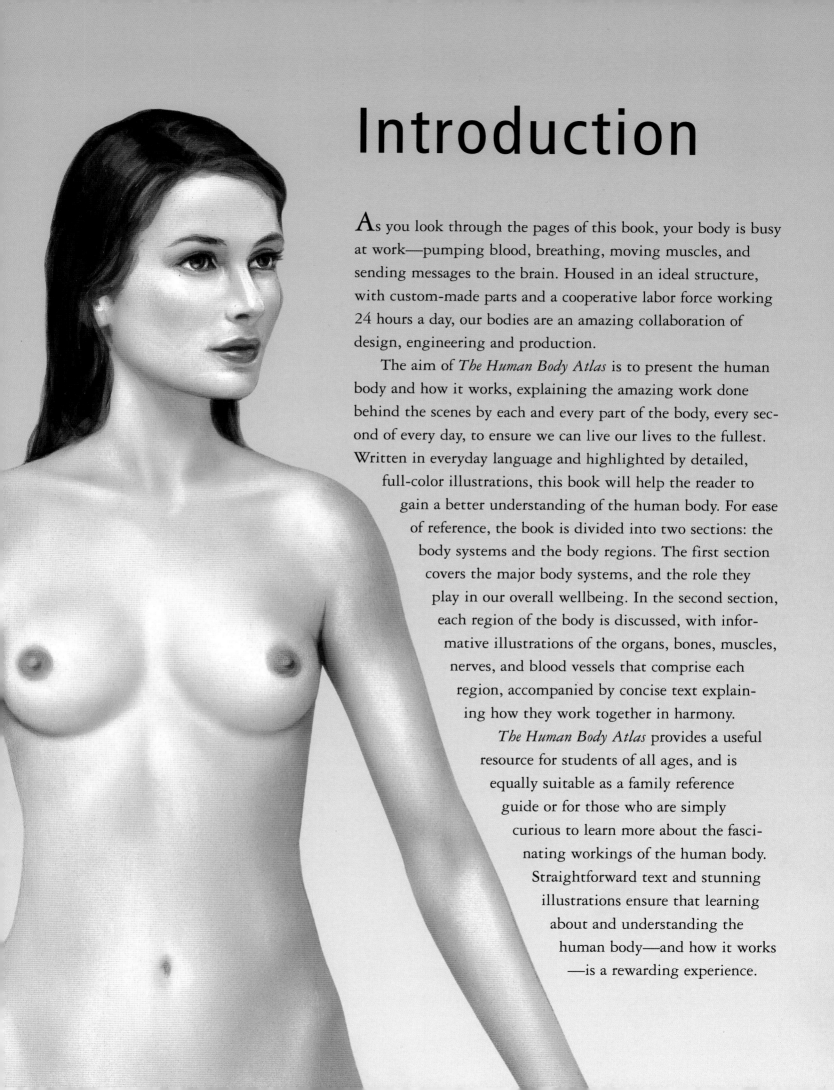

Introduction

As you look through the pages of this book, your body is busy at work—pumping blood, breathing, moving muscles, and sending messages to the brain. Housed in an ideal structure, with custom-made parts and a cooperative labor force working 24 hours a day, our bodies are an amazing collaboration of design, engineering and production.

The aim of *The Human Body Atlas* is to present the human body and how it works, explaining the amazing work done behind the scenes by each and every part of the body, every second of every day, to ensure we can live our lives to the fullest. Written in everyday language and highlighted by detailed, full-color illustrations, this book will help the reader to gain a better understanding of the human body. For ease of reference, the book is divided into two sections: the body systems and the body regions. The first section covers the major body systems, and the role they play in our overall wellbeing. In the second section, each region of the body is discussed, with informative illustrations of the organs, bones, muscles, nerves, and blood vessels that comprise each region, accompanied by concise text explaining how they work together in harmony.

The Human Body Atlas provides a useful resource for students of all ages, and is equally suitable as a family reference guide or for those who are simply curious to learn more about the fascinating workings of the human body. Straightforward text and stunning illustrations ensure that learning about and understanding the human body—and how it works—is a rewarding experience.

The Body Systems

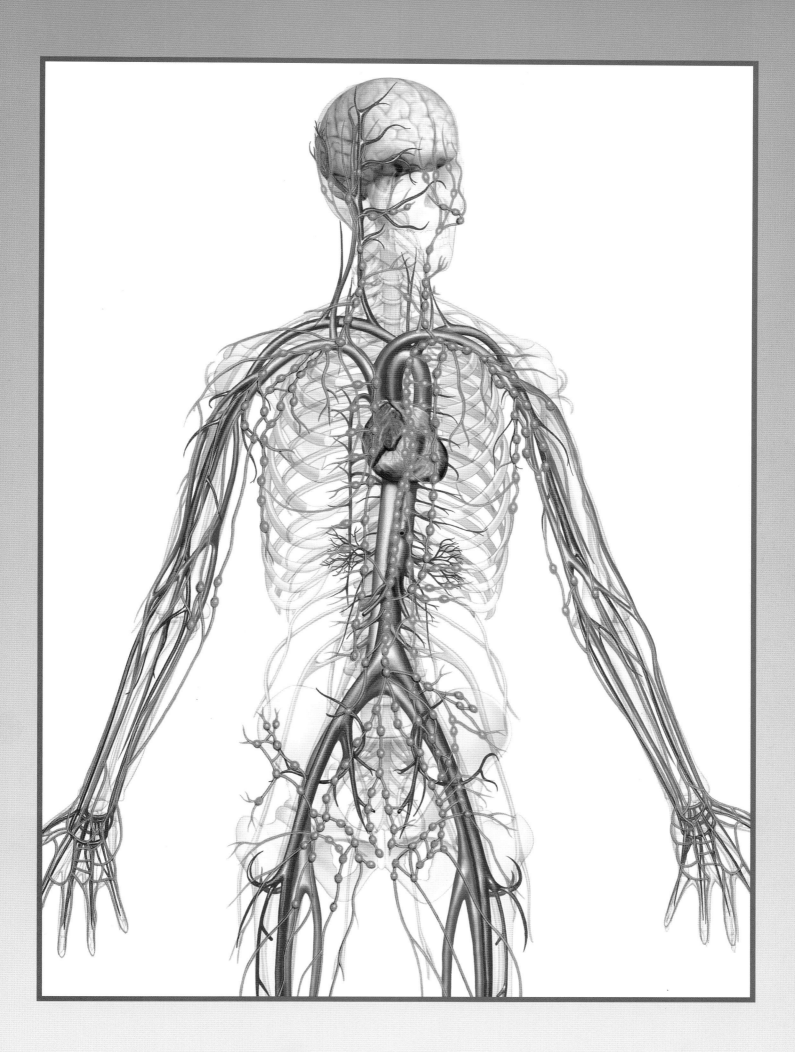

The Body Systems

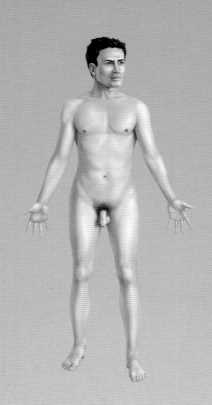

Muscular System

The muscles under our conscious control, the skeletal muscles, form the muscular system. Our physique is contoured by over 700 muscles, which account for around 60 percent of the mass of the body.

Skeletal System

The skeleton is the framework of the body and is usually described in two parts. The axial skeleton is comprised of the skull, spine, ribs and sternum; and the appendicular skeleton is comprised of the upper and lower limbs.

Digestive System

The digestive system is responsible for breaking down food into small, simple molecules for absorption and use as building blocks for cells.

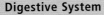

Respiratory System

The respiratory system exchanges carbon dioxide accumulated in the blood for oxygen in the airways. Inhaled air is taken into the lungs, and after gas exchange, carbon dioxide is exhaled.

Endocrine System

The endocrine glands secrete hormones that regulate growth, metabolism, sexual maturation and other important body functions. The main difference between the male and female systems is the male testes produces testosterone and the female ovaries produce estrogens and progesterone.

Circulatory System

There are two separate circulations within the body, the systemic circulation and the pulmonary circulation, with the heart providing the connecting link. The collaboration of the two systems ensures there is a constant supply of oxygen traveling through the body.

Nervous System

The nervous system is divided into a central nervous system (brain and spinal cord), and a peripheral nervous system (the remaining nerves in the body). The nervous system coordinates the body's response to messages received from the internal organs and the outside environment.

Lymphatic/Immune System

Following the course of the arteries and veins through the body, the lymphatic system carries interstitial fluid from cells and tissues back to the heart. Elements of the lymphatic/immune system search out and eliminate foreign bodies and invaders.

Urinary System

The urinary system is responsible for maintaining the correct levels of water and electrolytes in the body, filtering the blood and excreting waste or excess products. The urinary system comprises the kidneys, ureters, bladder and urethra.

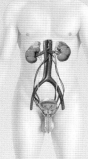

Reproductive System

The male reproductive system consists of the testes, the ductus deferens, the seminal vesicle, the prostate and the penis. The female reproductive system consists of the ovaries, fallopian tubes, uterus, vagina and external genitalia.

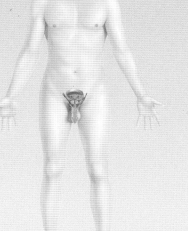
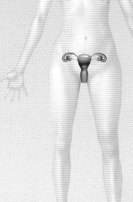

The Skeletal System

Generally described in two parts, the skeleton consists of the axial skeleton and the appendicular skeleton.

The Axial Skeleton

The axial skeleton comprises the skull, vertebral column, rib cage and sternum. The skull forms a protective shell for the brain and sensory organs, with the lower part of the skull giving form to the face. The base of the skull joins with the first vertebra of the spine, the atlas; the articulation of the two bones allows a range of movements, including nodding and sideways movement. Openings in the skull accommodate the eyes, nose, ears and mouth. The bones of the skull have unique joints, called sutures, which interlock and are then firmly held together with fibrous connective tissue.

The vertebral column (spine) is a tower of bones, called vertebrae, each separated by a cushioning pad of cartilage called the intervertebral disk. While individually each vertebra has extremely limited movement, collectively the vertebrae create a highly mobile unit. When required, the spine becomes a rigid unit, held firm by the back muscles, so that activities such as lifting can be performed.

The ribs of the thoracic cage join with the vertebrae at the back and encircle the heart and lungs. The first seven ribs, the true ribs, join at the front to the sternum. The next 3 ribs, the false ribs, join to one another and then attach to the last true rib. The remaining two ribs, the floating ribs, do not extend to the front.

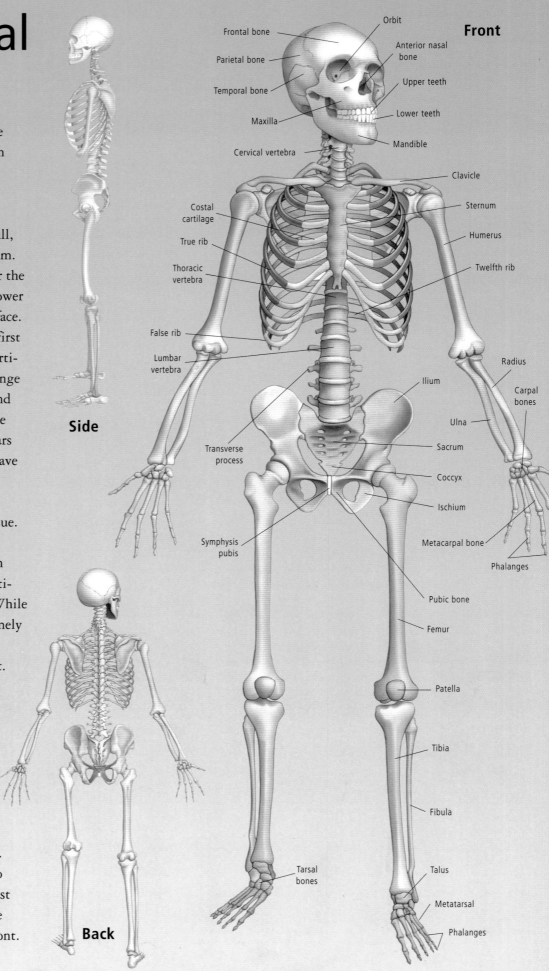

Side

Back

Front

Frontal bone
Orbit
Parietal bone
Anterior nasal bone
Temporal bone
Upper teeth
Maxilla
Lower teeth
Cervical vertebra
Mandible
Clavicle
Costal cartilage
Sternum
True rib
Humerus
Thoracic vertebra
Twelfth rib
False rib
Lumbar vertebra
Radius
Ilium
Carpal bones
Ulna
Transverse process
Sacrum
Coccyx
Ischium
Symphysis pubis
Metacarpal bone
Phalanges
Pubic bone
Femur
Patella
Tibia
Fibula
Tarsal bones
Talus
Metatarsal
Phalanges

The Appendicular Skeleton

The appendicular skeleton includes the limb bones of the arms and legs, and the girdles that connect them to the axial skeleton, the shoulder girdle and the pelvic girdle. The bones of the arm (humerus, radius and ulna) and the leg (femur, tibia and fibula) are all long bones.

With a similar structure, the bones of the hands and feet comprise 14 bones in the fingers and toes; the wrist has 8 bones, the ankle has 7; and there are 5 bones in both the palm of the hand and the sole of the foot.

Due to the load bearing requirements placed on the lower limbs, the pelvic girdle is joined to the axial skeleton at the sacroiliac joint, a relatively rigid joint, whereas the connection of the shoulder girdle is far less restricting, being mostly connected to the rib cage by muscle, its only stabilizing connection provided by the collar bone.

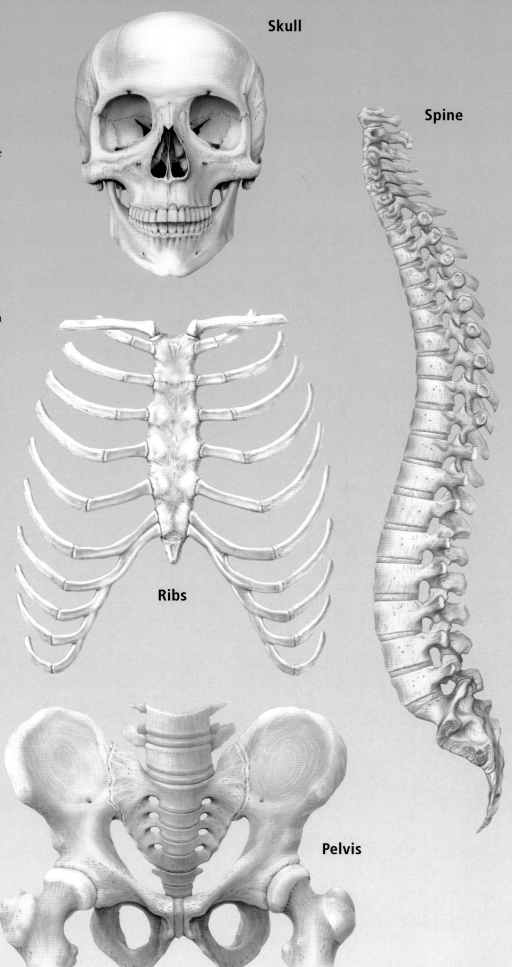

Skull

Spine

Ribs

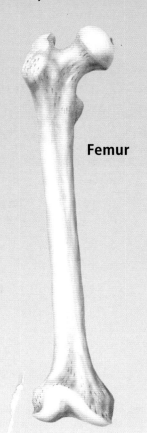

Femur

Pelvis

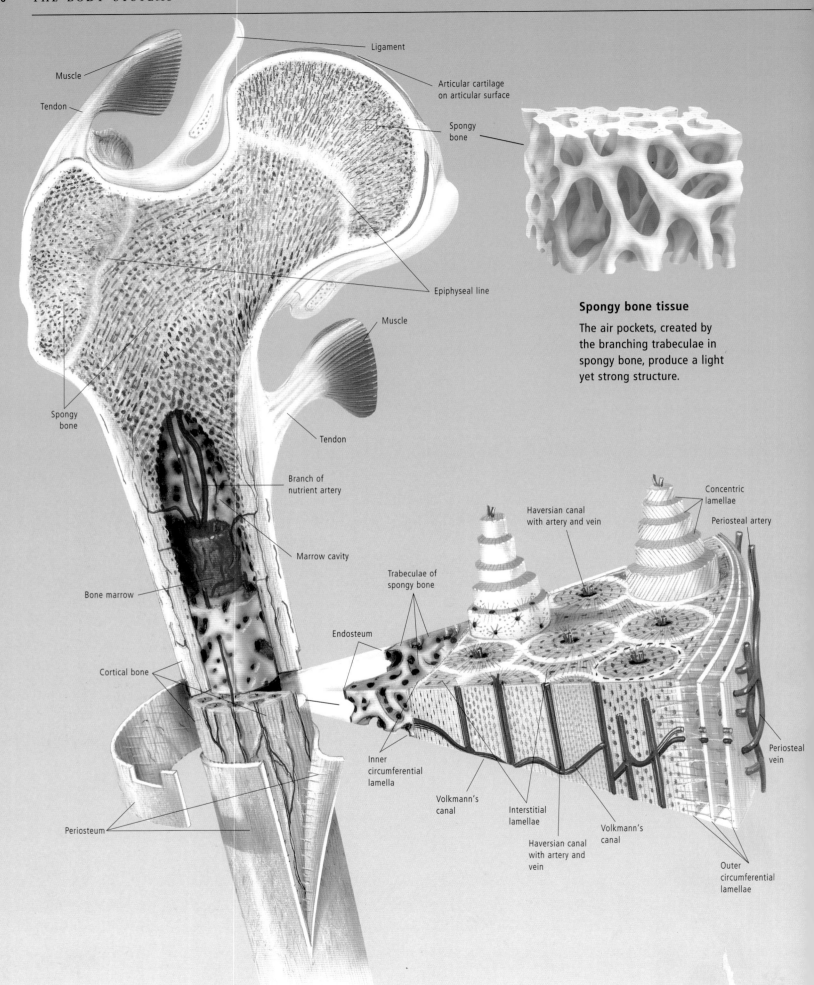

Muscle

Tendon

Ligament

Articular cartilage
on articular surface

Spongy
bone

Epiphyseal line

Muscle

Spongy
bone

Tendon

Branch of
nutrient artery

Marrow cavity

Bone marrow

Endosteum

Cortical bone

Periosteum

Spongy bone tissue

The air pockets, created by
the branching trabeculae in
spongy bone, produce a light
yet strong structure.

Concentric
lamellae

Haversian canal
with artery and vein

Periosteal artery

Trabeculae of
spongy bone

Inner
circumferential
lamella

Volkmann's
canal

Interstitial
lamellae

Volkmann's
canal

Haversian canal
with artery and
vein

Periosteal
vein

Outer
circumferential
lamellae

Bone Structure

The skeleton is comprised of bone—a rigid, calcified tissue, far stronger than its brittle appearance indicates. The bones of the skeleton provide a framework around the delicate internal organs, provide a point of attachment for muscles, and contain bone marrow where red blood cells are produced.

Long bones are identified by their long central shaft (diaphysis) and rounded ends (epiphyses). The exterior surface of the bone is surrounded by periosteum; this membrane layer supplies nerves and blood vessels to the bone. Periosteum covers all bone surfaces, except the joint surfaces, which have a cartilage covering.

Bone is composed of cells in a matrix. These cells perform various functions including bone tissue formation and bone tissue maintenance. The major components of the matrix include mineral salts, which provide hardness, and collagen fibers, which give strength.

Bone formation

(a) Bone grows in width as new bone is laid down in ridges.
(b) The ridges gradually enclose the blood vessels.
(c) More bone is laid down, reducing the space around the blood vessel.
(d) Eventually an osteon is formed.

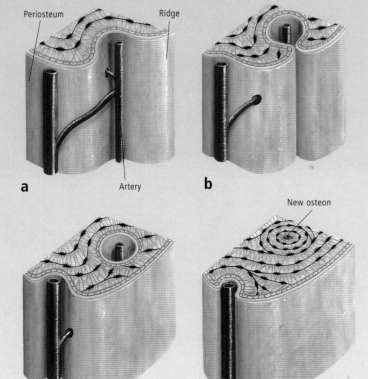

Periosteum   Ridge

Artery

a

b

New osteon

c

d

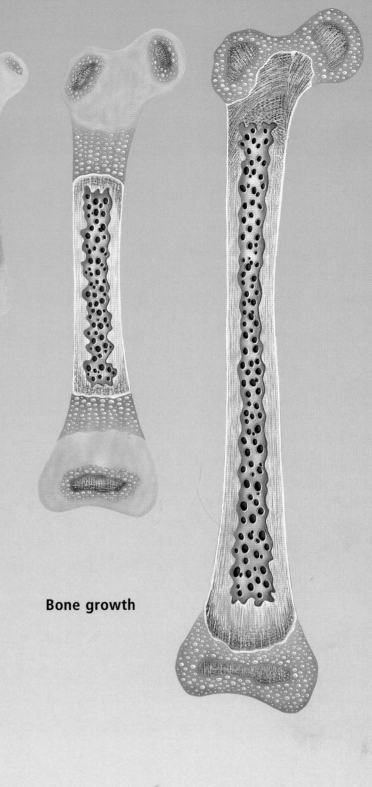

Bone growth

Bone Growth

Bone begins as cartilage in the developing embryo. By birth, ossification (development of bone) has taken place over most of the original cartilage base. Long bones then develop new centers for growth at either end of the shaft of the bone. Between the developing bone and the ossified bone is a layer of cartilage, the growth plate, which moves steadily away from the bone towards the ends until all cartilage is ossified.

Joints

Whenever we move, turn, bend or twist, our joints are involved. The point where two bones connect is an articulation, or joint. Exclusive to the skull are the joints known as sutures, which knit the bones of the skull together to form a stable, immobile joint, which is further strengthened by fibrous connective tissue.

Throughout the rest of the body are various mobile joint types: ball-and-socket, hinge, gliding, ellipsoidal, pivot and saddle joints. Varying degrees of mobility are achieved by the various joint types, with the ball-and-socket joint offering the greatest range of movement. Some joints, such as those joining at the front of the pelvis, are only slightly mobile, connected by a layer of cartilage and held firm by fibrous ligaments.

The surfaces of synovial joints are covered with smooth cartilage, and surrounded by a capsule lined with synovial membrane. The synovial membrane produces synovial fluid which acts to lubricate the joint, allowing smooth, virtually friction-free movement.

The joints are reinforced by ligaments, which prevent excessive movement. Often the arrangement of the connecting bone surfaces allows more than one type of joint movement—the elbow, for instance, includes a hinge joint and a pivot joint.

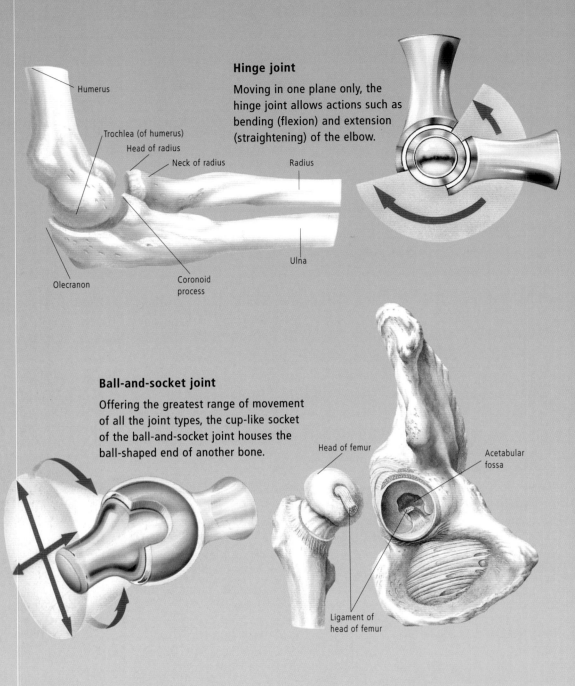

Hinge joint

Moving in one plane only, the hinge joint allows actions such as bending (flexion) and extension (straightening) of the elbow.

Humerus

Trochlea (of humerus)

Head of radius

Neck of radius

Radius

Ulna

Olecranon

Coronoid process

Ball-and-socket joint

Offering the greatest range of movement of all the joint types, the cup-like socket of the ball-and-socket joint houses the ball-shaped end of another bone.

Head of femur

Acetabular fossa

Ligament of head of femur

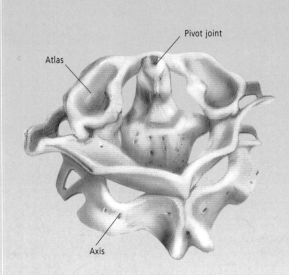

Pivot joint

Atlas

Axis

Pivot joint

The articulation between the first and second cervical vertebrae, the atlas and axis, allows rotational movement—this is known as a pivot joint.

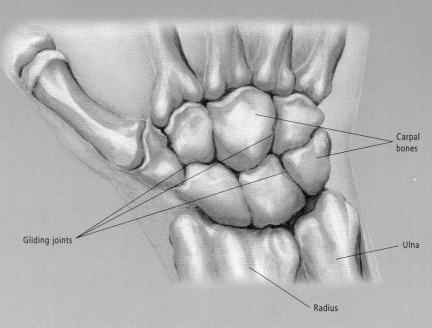

Carpal bones

Gliding joints

Ulna

Radius

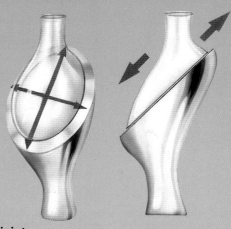

Gliding joint

Synovial fluid allows the bones of the gliding joint to slide across each other in limited movement.

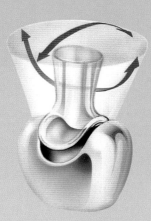

Saddle joint

A saddle joint allows sliding movement in two directions, offering almost as much movement as a ball-and-socket joint.

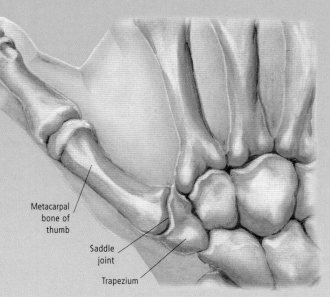

Metacarpal bone of thumb

Saddle joint

Trapezium

Ellipsoidal joint

Ulna

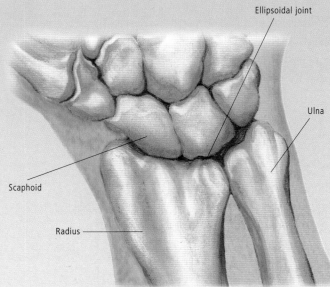

Scaphoid

Radius

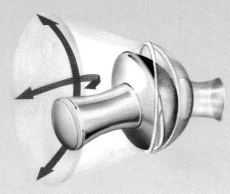

Ellipsoidal joint

An ellipsoidal joint allows movement in two directions, such as that which takes place at the wrist joint.

The Muscular System

We need muscles for every move we make. The muscular system springs into action to make our bodies move, in response to messages from the brain. Even when we are motionless, our muscular system is still at work, providing support to our skeletal frame. From the smallest twitch to the coordinated efforts of the breathing muscles, the muscular system powers all movement.

While the muscular system encompasses the skeletal muscle of the body—those muscles under our conscious control, known as the voluntary muscles—there are two other muscle types: cardiac muscle, found only in the heart, and smooth muscle, which is found in many of the internal organs. These two muscle types are known as involuntary muscles, and are controlled by the autonomic nervous system.

Beneath the skin's surface, layers of skeletal muscle cover our body, defining our physique. Usually attached at two points, skeletal

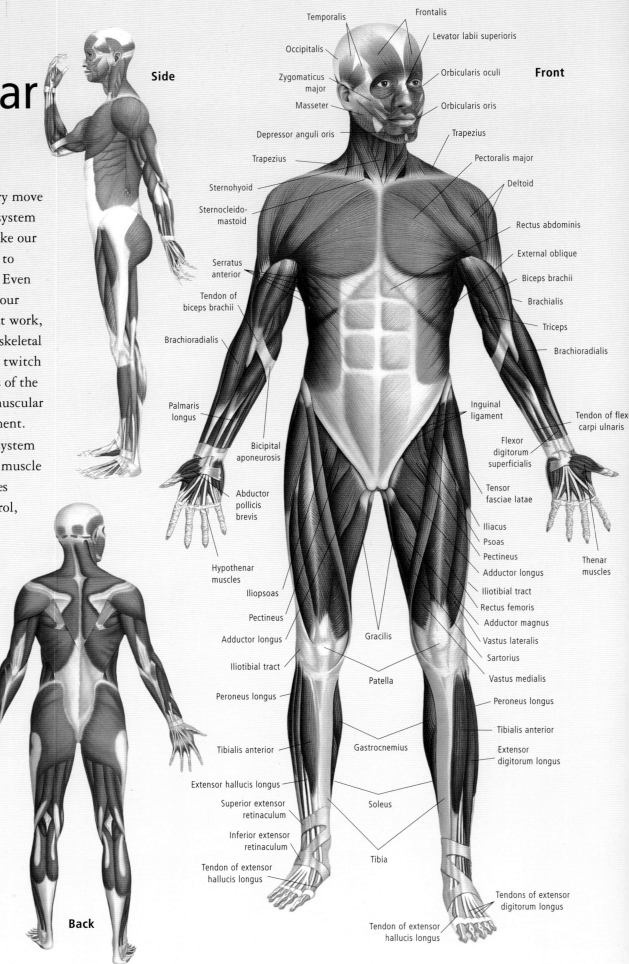

Side

Front

Temporalis
Frontalis
Occipitalis
Levator labii superioris
Zygomaticus major
Orbicularis oculi
Masseter
Orbicularis oris
Depressor anguli oris
Trapezius
Trapezius
Pectoralis major
Sternohyoid
Deltoid
Sternocleido-mastoid
Serratus anterior
Rectus abdominis
Tendon of biceps brachii
External oblique
Biceps brachii
Brachialis
Brachioradialis
Triceps
Brachioradialis
Palmaris longus
Inguinal ligament
Tendon of flex carpi ulnaris
Flexor digitorum superficialis
Bicipital aponeurosis
Tensor fasciae latae
Abductor pollicis brevis
Iliacus
Psoas
Pectineus
Thenar muscles
Hypothenar muscles
Adductor longus
Iliopsoas
Iliotibial tract
Pectineus
Rectus femoris
Adductor longus
Adductor magnus
Gracilis
Vastus lateralis
Iliotibial tract
Sartorius
Patella
Vastus medialis
Peroneus longus
Peroneus longus
Tibialis anterior
Tibialis anterior
Gastrocnemius
Extensor digitorum longus
Extensor hallucis longus
Soleus
Superior extensor retinaculum
Inferior extensor retinaculum
Tibia
Tendon of extensor hallucis longus
Tendons of extensor digitorum longus
Tendon of extensor hallucis longus

Back

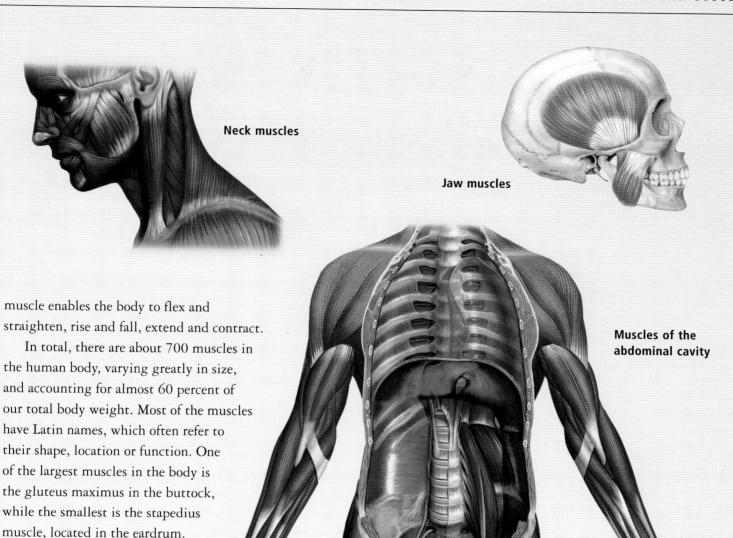

Neck muscles

Jaw muscles

Muscles of the abdominal cavity

muscle enables the body to flex and straighten, rise and fall, extend and contract.

In total, there are about 700 muscles in the human body, varying greatly in size, and accounting for almost 60 percent of our total body weight. Most of the muscles have Latin names, which often refer to their shape, location or function. One of the largest muscles in the body is the gluteus maximus in the buttock, while the smallest is the stapedius muscle, located in the eardrum.

Muscles are arranged in layers: superficial muscles lie close to the surface of the skin, while the deep muscle layers lie beneath, serving to protect the internal organs and body structures.

Muscles often work in pairs, with each of the muscles able to counteract the actions of the other. Muscles that work in this way are known as agonist and antagonist.

To produce movement, the muscles require a supply of oxygen and glucose; these nutrients are provided through an extensive network of blood vessels. The muscles also store some excess glucose, glycogen, as a fuel reserve.

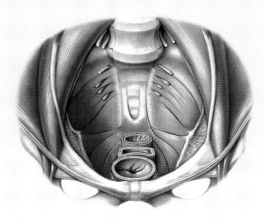

Pelvic floor muscles

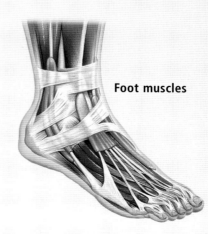

Foot muscles

Muscle Tissue

The three muscle types—skeletal muscle, smooth muscle and cardiac muscle—are constructed differently, as seen in the illustrations, requiring different compositions to suit their varying functions. Skeletal muscle and cardiac muscle are both striated (striped), while smooth muscle is non-striated.

Muscle Types

Skeletal muscle is made up of bundles of fibers. Wrapped in a layer of connective tissue, each of these bundles is called a fascicle, and their arrangement determines the function of the muscle. Most skeletal muscle has parallel fascicles, although the shape and appearance can vary depending on the function of the muscle. Muscles required to support organs and soft tissue have a network of interwoven fascicles. Fascicles can also be in a circular arrangement, allowing circular muscles to release or seal the entrance to major body passageways.

Muscle fiber microstructure

Muscle is comprised of bundles of fibers known as fascicles. Each fascicle contains strands called myofibrils. Each myofibril is comprised of thick and thin myofilaments. When prompted by messages from the brain, nerves trigger the interlocking action of myosin in the thick myofilaments with actin in the thin myofilaments, causing the muscle to contract; the connection is released and the myofilaments return to their original position when the muscle relaxes.

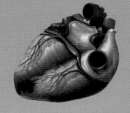

Cardiac muscle

The rhythmic movements of cardiac muscle are governed by the heart's natural pacemaker, the sinoatrial node, which is itself controlled by the autonomic nervous system.

Skeletal muscle

Controlled by the brain and spinal cord, skeletal muscle allows the body to move. Easy to detect below the surface of the skin, skeletal muscle, along with the skeletal framework, contours our physique.

Smooth muscle

Activated by the autonomic nervous system, smooth muscle is involuntary muscle found in the skin, blood vessels, and the reproductive and digestive systems.

Muscle types

Muscles can be classified based on their general shape—the arrangement of the fibers reflects the function of the muscle. Muscles used to move bone have fibers aligned in the same direction as the bone. Muscles required for the support of soft tissues have a latticework of overlapping fibers. Circular muscles, required for opening and closing of entrances, such as those found in the bowel or urinary tract, have fibers arranged in a circular pattern.

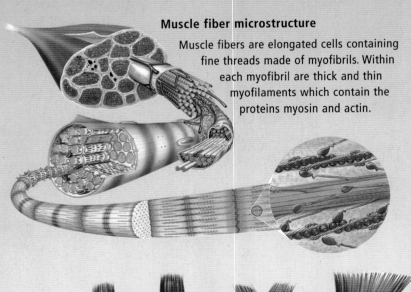

Muscle fiber microstructure

Muscle fibers are elongated cells containing fine threads made of myofibrils. Within each myofibril are thick and thin myofilaments which contain the proteins myosin and actin.

Quadrate   **Strap**   **Strap**   **Cruciate**   **Triangular**   **Multicaudal**   **Unipennate**   **Bipennate**   **Multipennate**

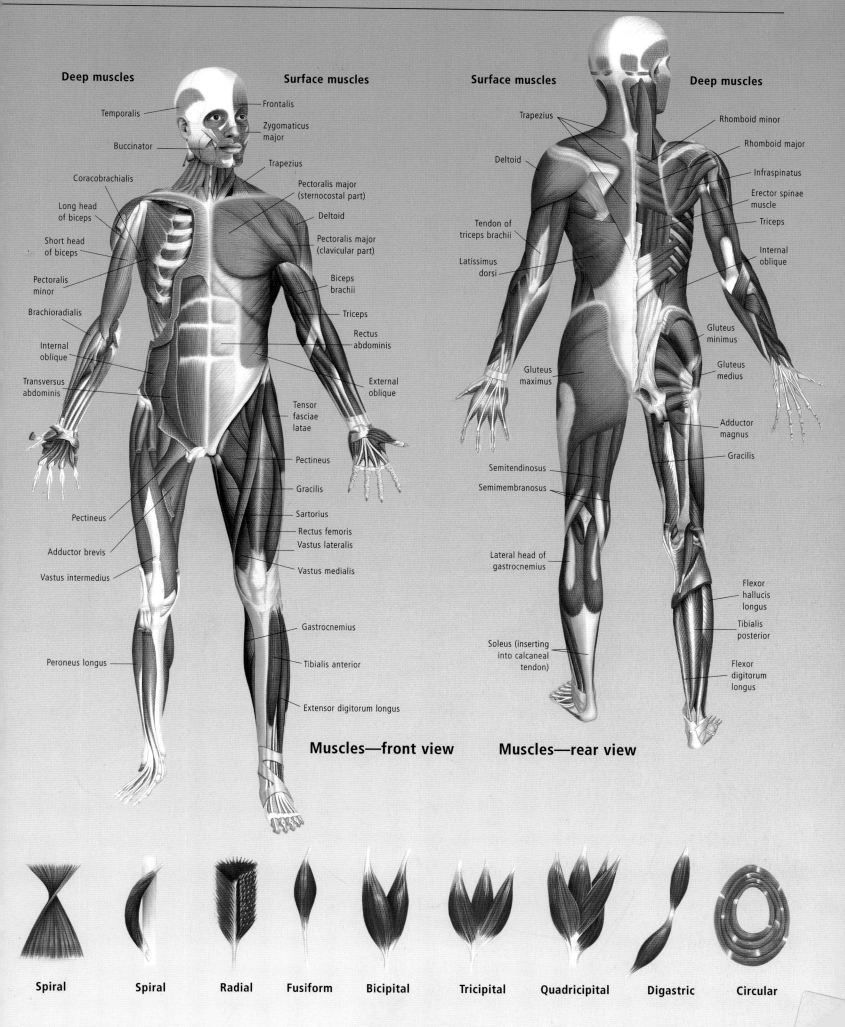

Deep muscles

Temporalis

Buccinator

Coracobrachialis

Long head of biceps

Short head of biceps

Pectoralis minor

Brachioradialis

Internal oblique

Transversus abdominis

Pectineus

Adductor brevis

Vastus intermedius

Peroneus longus

Surface muscles

Frontalis

Zygomaticus major

Trapezius

Pectoralis major (sternocostal part)

Deltoid

Pectoralis major (clavicular part)

Biceps brachii

Triceps

Rectus abdominis

External oblique

Tensor fasciae latae

Pectineus

Gracilis

Sartorius

Rectus femoris

Vastus lateralis

Vastus medialis

Gastrocnemius

Tibialis anterior

Extensor digitorum longus

Muscles—front view

Surface muscles

Trapezius

Deltoid

Tendon of triceps brachii

Latissimus dorsi

Gluteus maximus

Semitendinosus

Semimembranosus

Lateral head of gastrocnemius

Soleus (inserting into calcaneal tendon)

Deep muscles

Rhomboid minor

Rhomboid major

Infraspinatus

Erector spinae muscle

Triceps

Internal oblique

Gluteus minimus

Gluteus medius

Adductor magnus

Gracilis

Flexor hallucis longus

Tibialis posterior

Flexor digitorum longus

Muscles—rear view

Spiral **Spiral** **Radial** **Fusiform** **Bicipital** **Tricipital** **Quadricipital** **Digastric** **Circular**

The Nervous System

The central nervous system (CNS), comprised of the brain and the spinal cord, controls the motor and sensory functions of the body.

Weighing around 3 pounds (1.4 kilograms) and encased within the protective shell of the skull, the brain has a furrowed surface scored with deep fissures, ridges called gyri, and shallow folds called sulci. These fissures map out the various lobes of the cerebrum, the largest part of the brain. The lobes are the engine rooms for the processing of sensory and motor information. The brain is also vertically divided into two hemispheres—the left hemisphere controlling the right side of the body and the right hemisphere controlling the left side.

The cerebellum and brain stem form the remainder of the brain, and connect with the network of nerves in the spinal cord to complete the central nervous system. The spinal cord is the conduit for nerve impulses from the peripheral nervous system, conveying messages from the nerves to the brain, and relaying the response from the brain. The peripheral nervous system includes all the nerves distributed throughout the body.

The Functions of the Brain

The processing of motor and sensory information is performed by different regions within the lobes of the brain. The frontal lobe is responsible for thought processing and creativity. The parietal lobe processes feelings of pain, temperature, and touch. The occipital lobe governs sight, while the temporal lobe interprets sound.

The Cranial Nerves

There are 12 cranial nerves at the base of the brain, each responsible for controlling movement and for transmitting sensory information to various organs in the body. They also carry the nerve impulses for the five special senses—sight, smell, hearing, taste and balance, forwarding nerve impulses to various areas of the brain for processing.

The Brain Stem

The brain stem, comprising the midbrain, pons and medulla, is a vital relay station in the nervous system. The brain stem transmits messages to and from the brain along ascending and descending pathways respectively. Most of the cranial nerves arise in the brain stem; and processing of information about touch on the face, taste and hearing begins in the brain stem. Other important functions of the brain stem are the control of breathing, blood pressure and heart function.

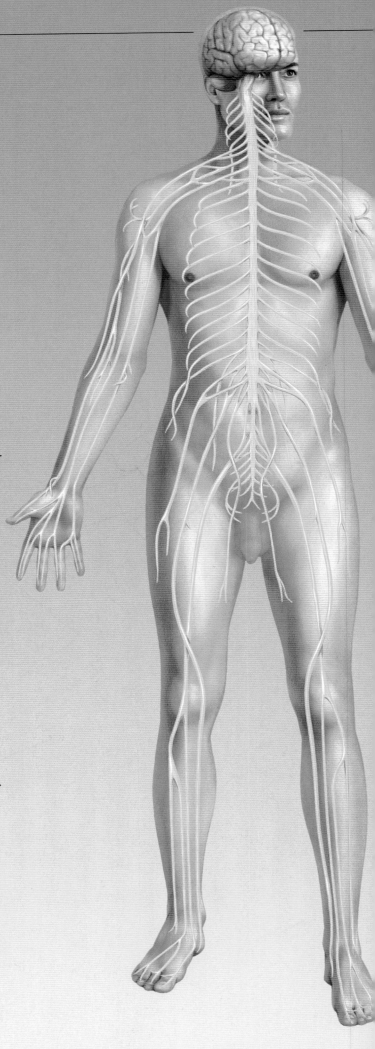

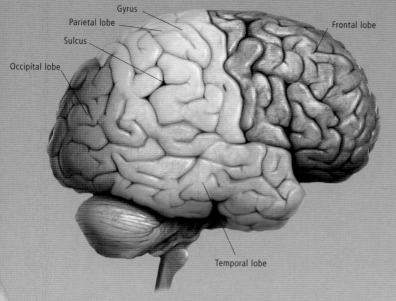

Gyrus
Parietal lobe
Sulcus
Occipital lobe
Frontal lobe
Temporal lobe

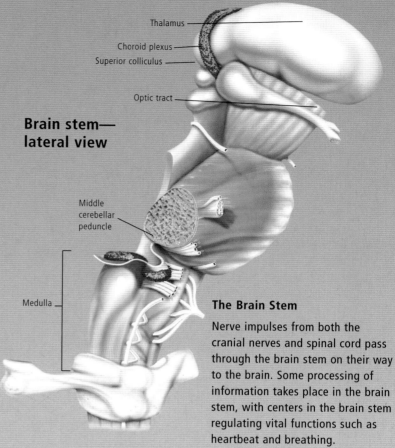

Brain stem—lateral view

Thalamus
Choroid plexus
Superior colliculus
Optic tract
Middle cerebellar peduncle
Medulla

The Lobes of the Brain

The brain is the major component of the central nervous system. The 4 lobes of the brain each have areas designed to process specific information. Areas of the frontal lobe are responsible for thought processing, problem solving, reasoning, and emotion; the temporal lobe is the center of auditory comprehension and memory; areas of the occipital lobe are responsible for visual interpretation; while the parietal lobe deals with feelings of pain, touch, and temperature regulation.

The Brain Stem

Nerve impulses from both the cranial nerves and spinal cord pass through the brain stem on their way to the brain. Some processing of information takes place in the brain stem, with centers in the brain stem regulating vital functions such as heartbeat and breathing.

Cranial nerves

The cranial nerves arise mainly from the brain stem. These 12 pairs of nerves provide innervation to various parts of the head and neck; the organs of the chest; and the upper part of the gastrointestinal tract. These nerves control movements of the face, tongue, eyes and throat (including skin, muscles and membrane) and receive sensory input from the special sense organs. The cranial nerves transmit information on sight, smell, hearing, balance and taste to the brain for processing.

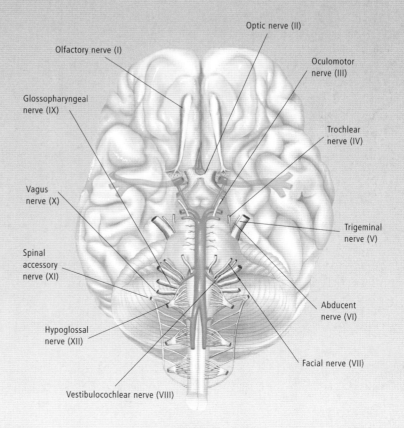

Optic nerve (II)
Olfactory nerve (I)
Oculomotor nerve (III)
Glossopharyngeal nerve (IX)
Trochlear nerve (IV)
Vagus nerve (X)
Trigeminal nerve (V)
Spinal accessory nerve (XI)
Abducent nerve (VI)
Hypoglossal nerve (XII)
Facial nerve (VII)
Vestibulocochlear nerve (VIII)

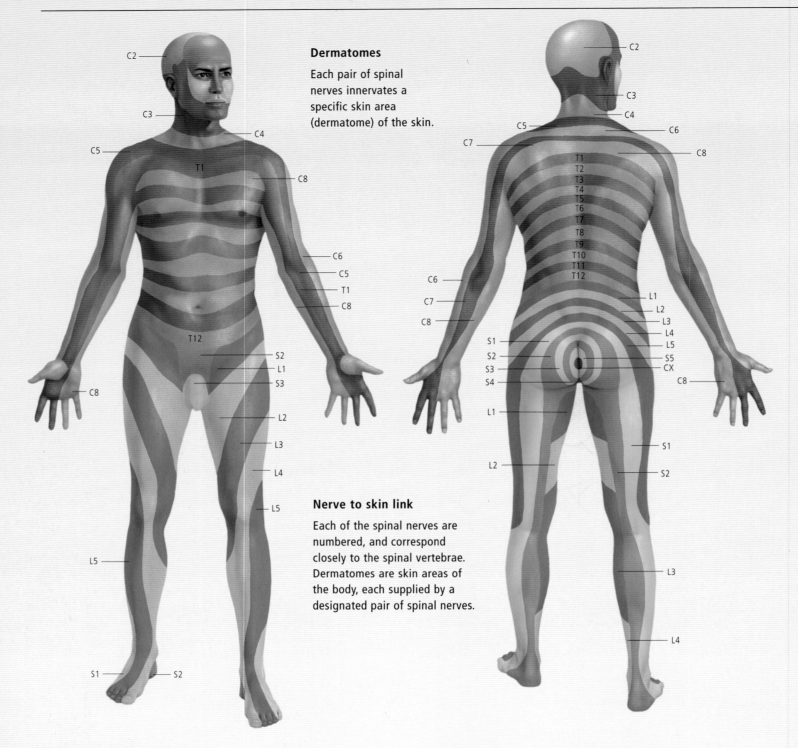

Dermatomes

Each pair of spinal nerves innervates a specific skin area (dermatome) of the skin.

Nerve to skin link

Each of the spinal nerves are numbered, and correspond closely to the spinal vertebrae. Dermatomes are skin areas of the body, each supplied by a designated pair of spinal nerves.

The Spinal Cord

A key element of the nervous system, the spinal cord acts as an intermediary between the peripheral nervous system and the brain. The cord has a central region of gray matter, which is divided into posterior (dorsal) and anterior (ventral) horns and an intermediate region. The gray matter receives and processes sensory information and sends signals to the muscles. The white matter surrounds the gray matter, and contains axons which transmit between the brain and the spinal cord.

There are 31 pairs of spinal nerves, which convey information about touch, pain, temperature, muscle tension and joint position, as well as branching out to service particular areas of the body. Each of the pairs of spinal nerves ends up innervating a different region of the skin called a dermatome.

Reflexes

Reflexes are our involuntary reaction to various stimuli. If we touch a hot surface, our hand is automatically pulled away by a reflex action. The nerve center in the spinal cord receives impulses from the nerves, and orders the appropriate response from a muscle or gland to counteract the action.

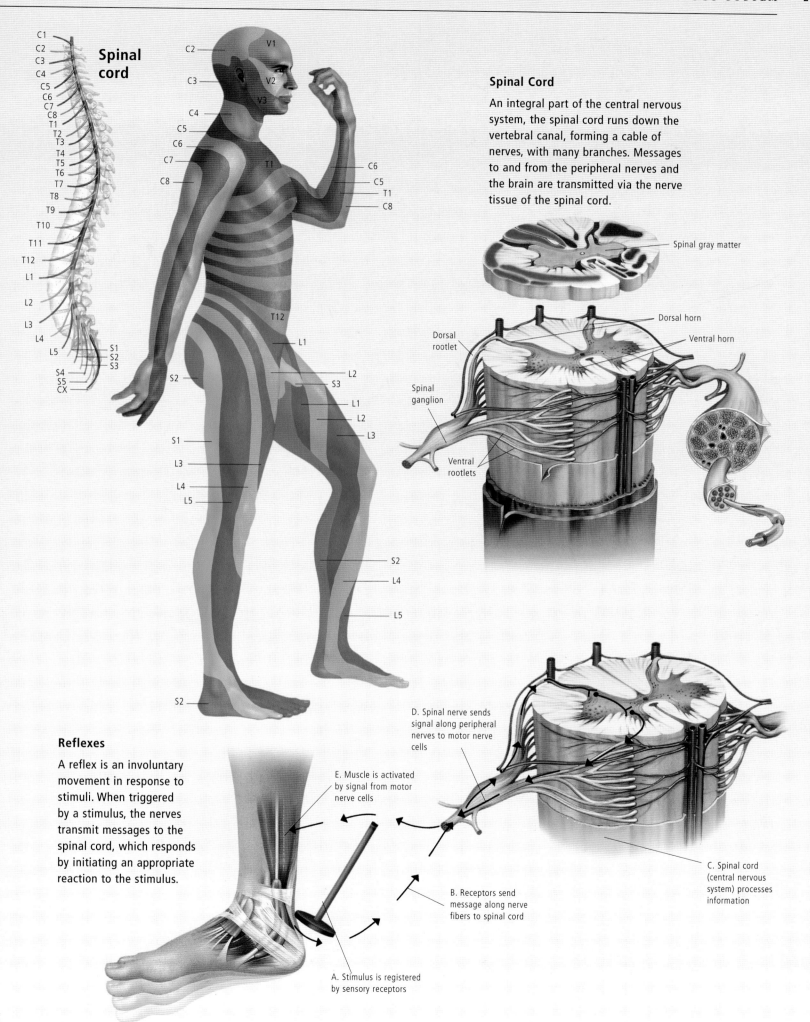

Spinal cord

C1
C2
C3
C4
C5
C6
C7
C8
T1
T2
T3
T4
T5
T6
T7
T8
T9
T10
T11
T12
L1
L2
L3
L4
L5
S1
S2
S3
S4
S5
CX

C2
V1
C3
V2
V3
C4
C5
C6
C7
C8
T1
C6
C5
T1
C8
T12
S2
L1
L2
S3
L1
L2
L3
S1
L3
L4
L5
S2
L4
L5
S2

Spinal Cord

An integral part of the central nervous system, the spinal cord runs down the vertebral canal, forming a cable of nerves, with many branches. Messages to and from the peripheral nerves and the brain are transmitted via the nerve tissue of the spinal cord.

Spinal gray matter

Dorsal horn
Ventral horn
Dorsal rootlet
Spinal ganglion
Ventral rootlets

Reflexes

A reflex is an involuntary movement in response to stimuli. When triggered by a stimulus, the nerves transmit messages to the spinal cord, which responds by initiating an appropriate reaction to the stimulus.

E. Muscle is activated by signal from motor nerve cells

D. Spinal nerve sends signal along peripheral nerves to motor nerve cells

C. Spinal cord (central nervous system) processes information

B. Receptors send message along nerve fibers to spinal cord

A. Stimulus is registered by sensory receptors

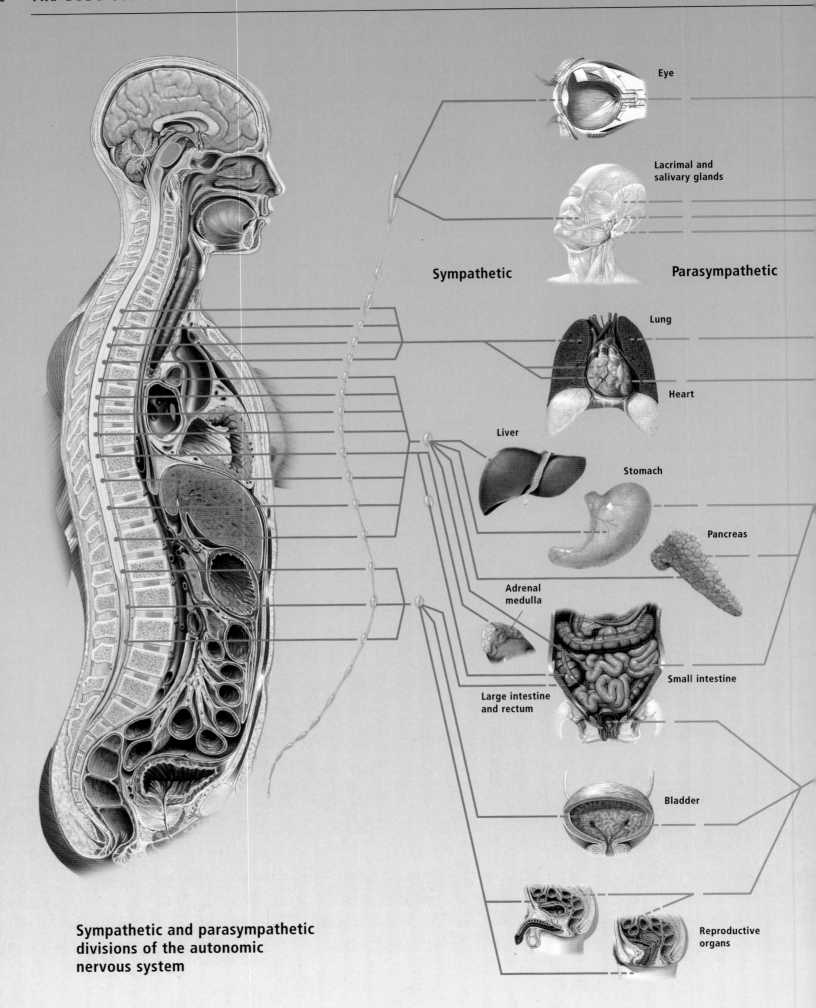

Eye

Lacrimal and
salivary glands

Sympathetic   **Parasympathetic**

Lung

Heart

Liver

Stomach

Pancreas

Adrenal
medulla

Small intestine

Large intestine
and rectum

Bladder

Reproductive
organs

**Sympathetic and parasympathetic
divisions of the autonomic
nervous system**

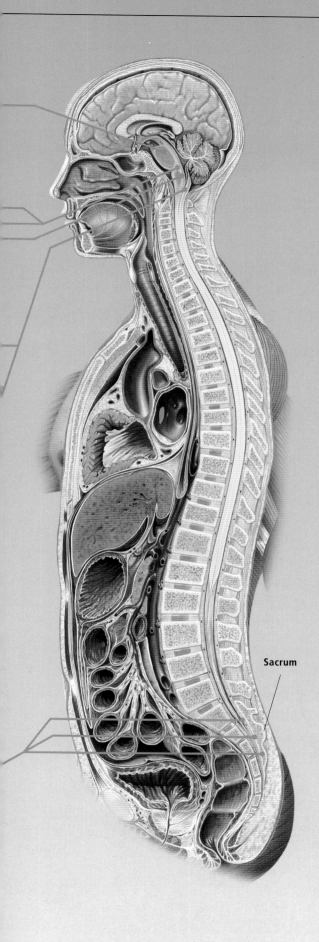

Sacrum

The Autonomic Nervous System

The autonomic nervous system is involved in controlling relatively automatic bodily functions. The autonomic nervous systems monitors internal conditions and keeps the brain and spinal cord informed of changes in the body. Regulating and controlling internal organs and processes, most of the actions of the autonomic nervous system occur without our being aware of them.

Divided into the sympathetic and parasympathetic division, each part is considered quite separate and performs different functions, although there is an overlap of the systems in some parts of the body.

The sympathetic division consists of nerve cells in the thoracic and lumbar levels of the spinal cord, as well as a long chain of nerves cells, called the sympathetic trunk, which lies along the spine. The sympathetic nervous system is often referred to as our "fight-or-flight" system, and springs into action in response to emergency situations. Reacting to an emergency, the autonomic nervous system can trigger responses such as a quickening heartbeat, increased breathing rate and enlarged pupils. Additionally, the sympathetic nervous system can cause changes such as increased blood pressure, a dry mouth, increased blood sugar levels, dilation of the small airways of the lung, and increased blood flow to the muscles. All of these measures increase the individual's ability to cope with the emergency. The sympathetic division is also responsible for regulating body temperature.

The parasympathetic division is most active when the body is not under threat and when it is largely at rest. Its main function is to conserve energy, restoring the internal body state to normal by promoting digestion and eliminating urine and feces from the body.

The autonomic nervous system

The thoracic, abdominal and pelvic organs and tissues come under the dual influence of the sympathetic and parasympathetic divisions of the autonomic nervous system, with the two systems working together to ensure a relatively constant state of activity. Autonomic innervation of the muscles of the eye, salivary glands, sweat glands, and muscles in the skin is also governed by the autonomic nervous system.

The Circulatory System

The blood vessels and heart form the circulatory system, pumping blood through their complete circuit, and carrying oxygen to the organs and tissues of the body. The heart pumps oxygen-rich blood through the body, distributing it through arteries and capillaries, and supplying oxygen and nutrients vital to the functioning of the internal organs and tissues. After exchanging gases with surrounding tissues, the blood, now oxygen-depleted and laden with carbon dioxide, returns to the heart through the veins. This is known as the systemic circulation, one of two separate circulations. The other, known as pulmonary circulation, transfers blood returning to the heart, along the pulmonary trunk and arteries, to the lungs for oxygen replenishment and elimination of carbon dioxide. This renewed blood is then returned to the heart, returning to the systemic circulation, ready for another circuit of the body.

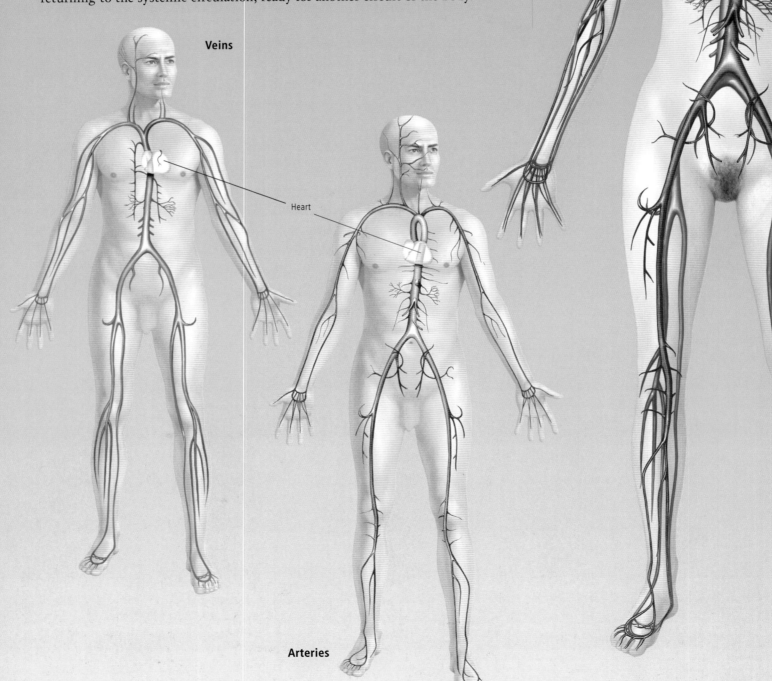

Veins

Heart

Arteries

The Heart

The fist-sized heart is a muscular pump, rhythmically pulsing to maintain the supply of blood to the body in a continuous process, with the ventricles of the heart playing an integral part. Blood from the left ventricle delivers oxygen and nutrients to the whole body via the systemic circulation. Once a circuit of the body has been completed, blood from the right ventricle collects oxygen from the lungs via the pulmonary circulation. The flow of blood through both the systemic and pulmonary circulations is controlled by the valves of the heart. The mitral valve and tricuspid valves (the atrioventricular valves), working in unison with the aortic and pulmonary valves (the semilunar valves), control blood flow, opening and closing in precise synchronized movements and, when closed, create a perfect seal, preventing blood from flowing back into the heart. When the ventricles contract (ventricular systole), the aortic and pulmonary valves open to allow blood to be pumped into the pulmonary and systemic circulatory systems, while the mitral and tricuspid valves remain closed. When the ventricles dilate (ventricular diastole), the aortic and pulmonary valves close, while the tricuspid and mitral valves open to allow blood to pass from the atria into the ventricles.

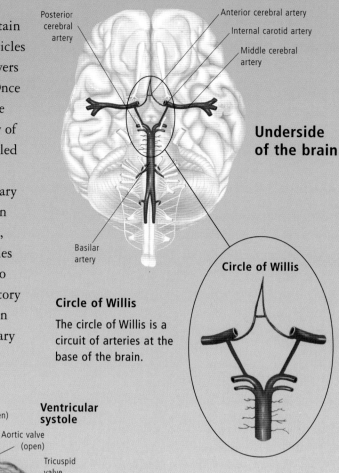

Underside of the brain

Posterior cerebral artery
Anterior cerebral artery
Internal carotid artery
Middle cerebral artery
Basilar artery

Circle of Willis

Circle of Willis

The circle of Willis is a circuit of arteries at the base of the brain.

Ventricular diastole

Pulmonary valve (closed)
Aortic valve (closed)
Tricuspid valve (open)
Mitral valve (open)

Ventricular systole

Pulmonary valve (open)
Aortic valve (open)
Tricuspid valve (closed)
Mitral valve (closed)

Heart valves

The coordinated movements of the valves of the heart ensure blood flow to the body is maintained. The mitral and tricuspid valves open to allow blood from the atrium to enter the ventricle, while the aortic and pulmonary valves remain closed. When the triscuspid and mitral valves close, the aortic and pulmonary valves open to release blood from the ventricles into the systemic and pulmonary circulations.

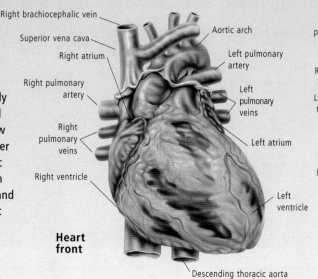

Right brachiocephalic vein
Superior vena cava
Right atrium
Right pulmonary artery
Right pulmonary veins
Right ventricle
Aortic arch
Left pulmonary artery
Left pulmonary veins
Left atrium
Left ventricle
Descending thoracic aorta

Heart front

Heart front— cross-section

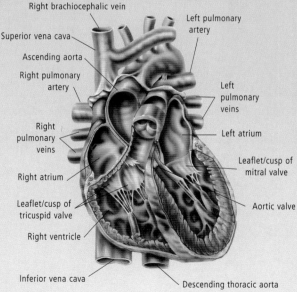

Right brachiocephalic vein
Superior vena cava
Ascending aorta
Right pulmonary artery
Right pulmonary veins
Right atrium
Leaflet/cusp of tricuspid valve
Right ventricle
Inferior vena cava
Left pulmonary artery
Left pulmonary veins
Left atrium
Leaflet/cusp of mitral valve
Aortic valve
Descending thoracic aorta

Heart

The heart is divided into four chambers: the left and right atria and the left and right ventricles. It pumps blood out from the left and right ventricles, and collects returned blood into its left and right atria.

Blood

Blood is a suspension of red and white blood cells, platelets, proteins and chemicals in a fluid called plasma. Plasma generally constitutes half the volume of blood, while the red and white blood cells account for the other half. Blood carries oxygen and essential nutrients from the lungs and digestive tract to other parts of the body, and carries waste products to organs such as the kidneys and lungs for elimination.

Produced in the bone marrow, the red blood cells (erythrocytes) carry oxygen to the body tissues, exchanging oxygen for carbon dioxide, which they then take to the lungs where it is exhaled. The white blood cells each play a role in the immune system; some produce antibodies to fight injury and infection, while others engulf bacteria and foreign invaders. Platelets are fragments of large cells called megakaryocytes. They are involved in the clotting process.

Blood vessels

A massive network of blood vessels circulate blood through the body. Blood pumped from the heart to the aorta, a large elastic artery, is then circulated to arteries, which subdivide many times, until they become capillaries. It is in the capillaries that oxygen and nutrients is exchanged for carbon dioxide by the cells. After leaving the limbs and organs, blood is channeled into veins of increasing size, returning eventually to the heart to repeat the cycle.

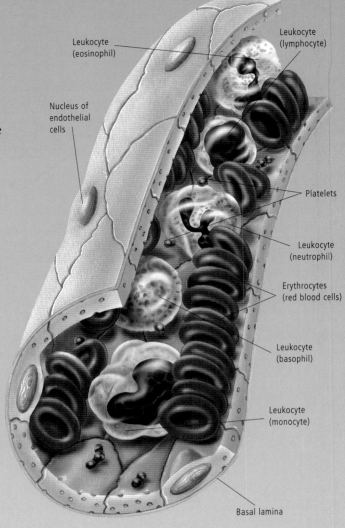

Leukocyte (eosinophil)

Leukocyte (lymphocyte)

Nucleus of endothelial cells

Platelets

Leukocyte (neutrophil)

Erythrocytes (red blood cells)

Leukocyte (basophil)

Leukocyte (monocyte)

Basal lamina

White blood cells

Monocyte
Monocytes become macrophages after circulating in the blood for 1–2 days.

Macrophage
Macrophages engulf foreign organisms and debris to fight infection.

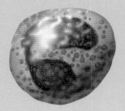

Eosinophil
Enzymes released by eosinophils cause allergic reactions and kill some parasites.

Neutrophil
Neutrophils engulf and destroy microorganisms, defending the body against bacterial invasions.

Basophil
Substances released by basophils increase the body's response to invading allergens.

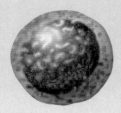

Lymphocyte
There are 3 types of lymphocyte. Natural killer cells and T cells attack foreign invaders directly; B cells make antibodies.

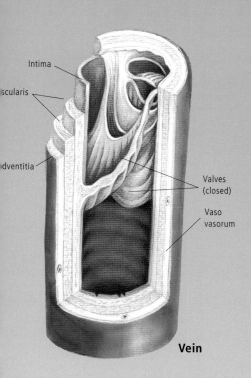

Intima

scularis

dventitia

Valves
(closed)

Vaso
vasorum

Vein

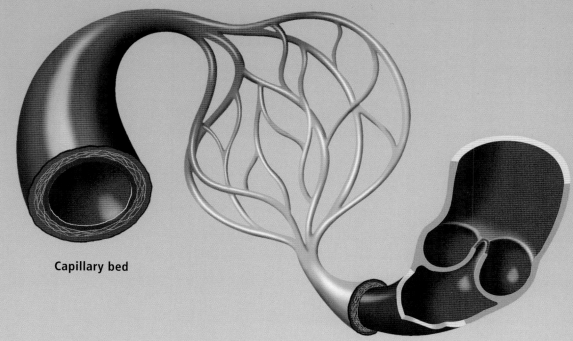

Capillary bed

Blood vessels

Several different types of blood vessels
service the body. Arteries are thick-walled
and flexible, while veins are thin-walled
and some contain valves to prevent the
backflow of blood. The arteries and veins
branch repeatedly to form thin-walled
capillaries, across which oxygen and
nutrients are exchanged for carbon
dioxide by the body cells.

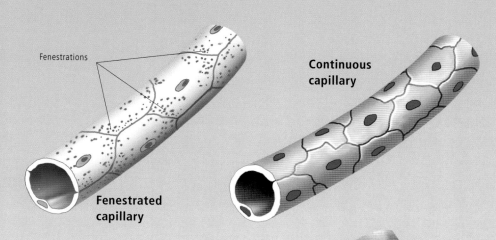

Fenestrations

**Continuous
capillary**

**Fenestrated
capillary**

Platelets

The tiny platelets trigger the
clotting system, preventing
bruising and bleeding.

Red blood cell production

Red blood cells are formed in the
bone marrow. The rate of production
is extremely high, which is necessary
considering the huge numbers of
red blood cells in the body. Ageing
red blood cells are filtered out of
the system by the spleen, which
also stores red blood cells,
for release when needed.

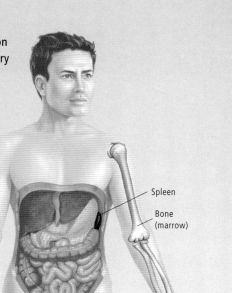

Spleen

Bone
(marrow)

Red blood cells

The shape of red blood cells enables them
to maneuver through small capillaries and
to stack together to facilitate blood flow.

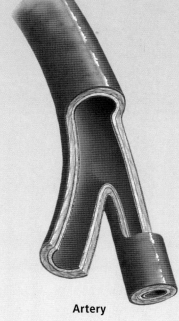

Artery

The Lymphatic/ Immune System

The lymphatic system is a network of vessels, nodes and organs serving a dual purpose in the body.

Firstly, the lymphatic system is responsible for draining interstitial fluid back to the circulatory system. Interstitial fluid filters out from blood vessels to bathe the body's tisues. This fluid is then absorbed into the lymphatic vessels (along with tissue and cellular wastes), cleaned and returned to the blood vessels, in a continual recycling process. The fluid flowing through the lymphatic vessels is known as lymph, and contains water, protein molecules, salts, glucose, urea and disease-fighting white blood cells.

The other important role of the lymphatic system is that of combating disease. The lymphatic system is packed with specialized white blood cells which use various methods to eradicate foreign bodies or invaders, such as bacteria, viruses and cancer cells.

Lymph Vessels

The lymphatic system is a one-way system of lymph vessels. Beginning at blind-ended capillaries, the vessels gradually increase in size, and pass through aggregates of lymphoid tissue or lymph nodes, finally converging into the large lymph trunks which drain into the right lymphatic duct and the thoracic duct, which, in turn, empty into the veins of the circulatory system.

The lymph vessels contain valves to prevent the backflow of lymph fluid, and with no pumping system, rely on movements of skeletal muscle to pump lymph back to the heart.

DID YOU KNOW?

The average adult is made up of around 100 trillion cells, of which about 1 trillion cells make up the immune system.

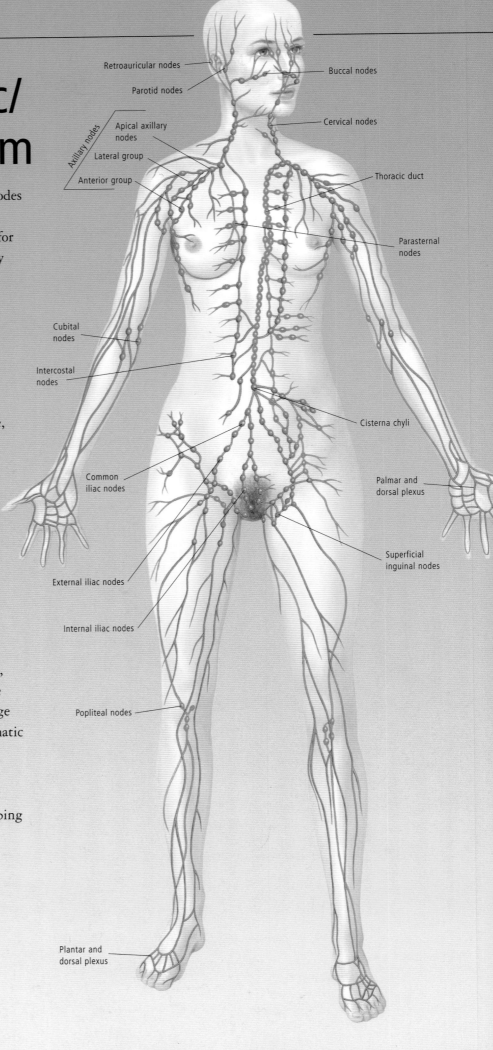

Retroauricular nodes

Buccal nodes

Parotid nodes

Cervical nodes

Axillary nodes

Apical axillary nodes

Lateral group

Thoracic duct

Anterior group

Cubital nodes

Parasternal nodes

Intercostal nodes

Cisterna chyli

Common iliac nodes

Palmar and dorsal plexus

Superficial inguinal nodes

External iliac nodes

Internal iliac nodes

Popliteal nodes

Plantar and dorsal plexus

Lymph nodes

The lymph vessels pass through lymph nodes, which manufacture lymphocytes, with each node connected to incoming and outgoing lymphatic vessels. Each lymph node consists of a mass of lymphatic tissue surrounded by a fibrous capsule. The lymph nodes are usually grouped together; concentrations of lymph nodes which serve the head and limbs are found in the lower jaw and neck, the armpit (axilla), and the groin; lymph from the internal organs of the thorax and abdomen drains into chains of lymph nodes along major arteries and the aorta. Lacteals, important in the absorption of fats, are lymphatic vessels in the walls of the digestive system which collect large molecules and lipids (chyle) extracted from food.

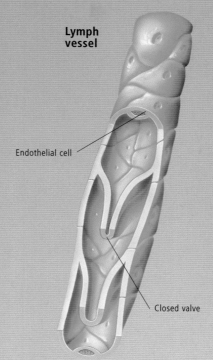

Lymph vessel

Endothelial cell

Closed valve

Lymph circulation

The lymph vessels contain numerous valves which prevent the backflow of lymph. The lymph is returned to the general circulation by two large lymphatic vessels which empty into the large veins at the base of the neck.

Lymph node

The lymph nodes are a mass of lymphatic tissue enclosed in a fibrous capsule. Each lymph node is connected to an incoming (afferent) and outgoing (efferent) lymph vessel.

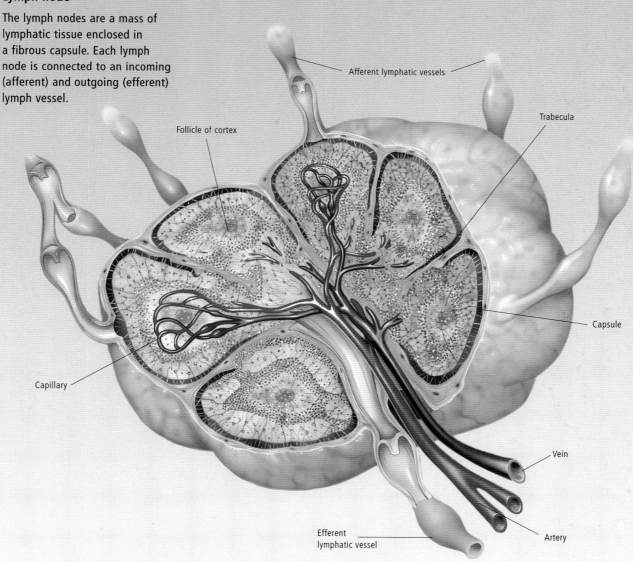

Afferent lymphatic vessels

Follicle of cortex

Trabecula

Capsule

Capillary

Vein

Efferent lymphatic vessel

Artery

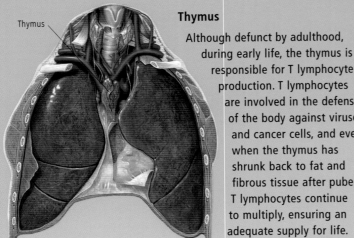

Thymus

Thymus

Although defunct by adulthood, during early life, the thymus is responsible for T lymphocyte production. T lymphocytes are involved in the defense of the body against viruses and cancer cells, and even when the thymus has shrunk back to fat and fibrous tissue after puberty, T lymphocytes continue to multiply, ensuring an adequate supply for life.

Spleen

The largest concentration of lymphatic tissue in the body, the spleen acts as a filter, removing ageing blood cells from the circulation, and is a center for lymphocyte production and storage.

Lymphocyte

There are three types of lymphocyte. Natural killer cells and T cells attack foreign invaders directly; B cells make antibodies.

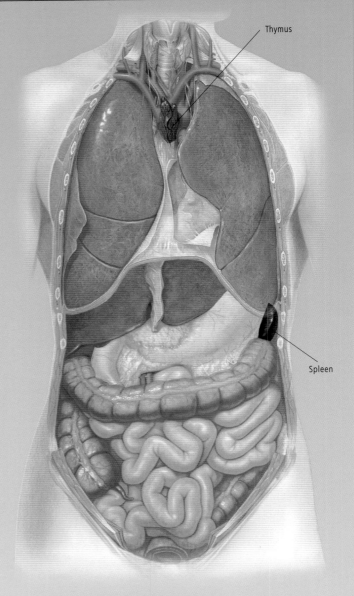

Thymus

Spleen

The organs of the lymph system

The thymus, spleen and lymphoid tissue found in the linings of the respiratory, urogenital and reproductive organs are the main organs associated with the lymphatic system.

The Lymph Organs

The lymphoid organs include the thymus, the spleen and mucosa-associated lymphoid tissue.

The thymus, found in the upper part of the chest between the heart and the sternum, is the first lymphoid organ to develop in the embryo. It continues to grow, reaching $1-1^{1}/_{2}$ ounces (30–40 grams) at puberty, after which time it gradually reduces to a fatty fibrous tissue, weighing only about $^{1}/_{2}$ ounce (14 grams). The thymus secretes hormones, which it uses to manufacture mature T lymphocytes. In immune response, these specialized lymphocytes recognize foreign-body antigens. Although the thymus withers away after puberty, T cells continue to multiply, ensuring a continuous supply for the body's requirements throughout life.

The spleen, found in the left side of the abdomen beneath the diaphragm, has the largest concentration of lymphatic tissue in the body. The spleen filters blood through an extensive network of capillaries and sinuses, called the red pulp. Concentrations of lymphocytes in the spleen are called the white pulp. Lymphocytes are produced and stored in the spleen.

The mucosa-associated lymphoid tissue is found in the linings of the respiratory, urogenital and digestive tracts. These tissues contain B and T lymphocytes, presenting a line of defense in those cavities of the body exposed to foreign invaders from the external environment.

Lymphocytes

Lymphocytes are crucial to the body's immune system. There are 3 major types of lymphocytes: B lymphocytes, T lymphocytes and natural killer cells (NK cells). B lymphocytes are produced in the bone marrow, while T lymphocytes are produced in the thymus. Lymphocytes either react to specific antigens, stimulating the production of antibodies, or engulf invading foreign cells.

Humoral immune response

B lymphocytes produce antibodies to help identify and eliminate invading antigens (carried by bacteria or viruses). They are helped by circulating T lymphocytes and macrophages.
(a) Virus particles invade tissue through surface cells and multiply.
(b) Virus particles are consumed by macrophages.
(c) The macrophages present antigens to circulating T lymphocytes. These recruit more T and B lymphocytes to help defend the body.
(d) B lymphocytes divide into plasma B cells, which make antibodies specific to the invading virus, and memory B cells.
(e) The circulating antibodies attach onto the virus particles.
(f) Macrophages recognize and engulf the virus, saving the body from infection.

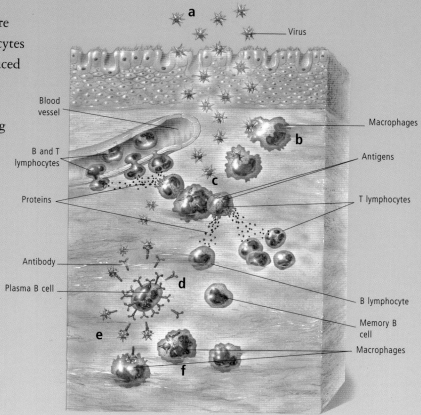

Allergic reaction

An allergic response is a reaction by the body to foreign irritants called allergens.
(a) Allergens enter the body causing the plasma B cells to produce antibodies.
(b) The antibodies attach to mast cells circulating in the body's tissues.
(c) When allergens again enter the body they are captured by the antibodies on the mast cells.
(d) The mast cells respond by releasing histamine, which produces the irritating symptoms of the allergy.

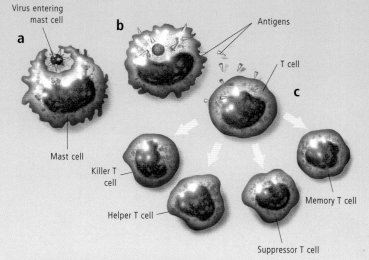

Cell-mediated immune response

T lymphocytes are responsible for the delayed action of the cell-mediated response.
(a) Circulating mast cells ingest invading virus.
(b) Mast cells process the virus and present antigens to T cells.
(c) The T cells produce clones which each play a special role in the immune response: memory T cells remember the invading antigen for future attacks; helper T cells recruit B and T cells to the site of antigen attack; suppressor T cells inhibit the action of B and T cells; and killer T cells attach onto invading antigens and destroy them.

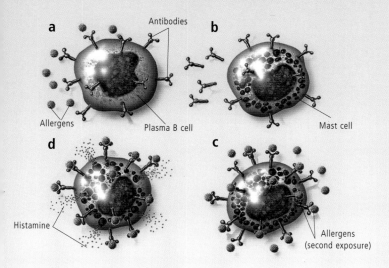

The Digestive System

The digestive system extracts nutrients from food, breaking it down into small molecules for absorption into the body. All the nutrients—proteins, fats, carbohydrates, vitamins and minerals—essential to maintain vital body functions, are absorbed into the body via the gastrointestinal tract (alimentary canal), where food is progressively broken down.

The Alimentary Canal

The muscles of the mouth enable food to be moved around, while the teeth grind, chew and shred. Saliva produced by the salivary glands moistens the food to make chewing and swallowing easier. The chewed food, formed into a mass called a bolus, is pushed to the back of the mouth, and from there muscular movement pushes it into the esophagus. The smooth muscle of the esophagus uses wave-like movements to propel the food to the stomach.

The stomach churns and processes the food, using muscular contractions to mix the digested food with the gastric juices of hydrochloric acid and pepsin; the resultant mixture is known as chyme. The opening of the stomach into the first section of the small intestine, the duodenum, is controlled by a thick bundle of circular muscle fibers, the pylorus. The pylorus gradually releases the chyme into the duodenum, where absorption of nutrients begins. Bile from the liver and enzymes from the pancreas empty into the duodenum. The lining of both the duodenum and jejunum is heavily folded, with each fold covered in tiny projections (villi), creating a large surface area for absorption of nutrients. Most of the nutrients from the digested food are absorbed in the duodenum and jejunum. Small molecules enter the blood capillaries, while the larger molecules enter the lymphatic channels (lacteals).

The final section of the small intestines, the ileum, absorbs Vitamin B_{12} and is responsible for returning bile acids back to the liver.

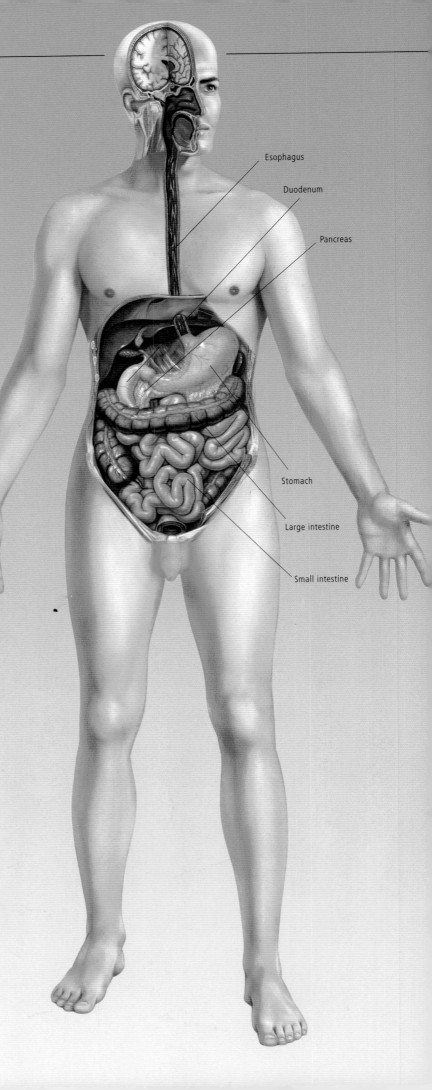

Esophagus

Duodenum

Pancreas

Stomach

Large intestine

Small intestine

Peristalsis

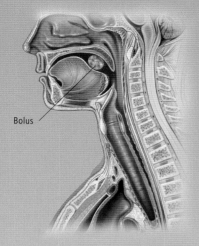 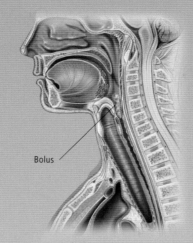 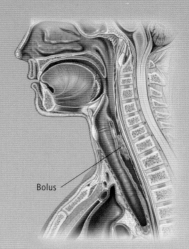

Bolus

Bolus

Bolus

Swallowing is under voluntary control; when swallowing begins, the tongue propels the bolus of food into the pharynx. Coordinated voluntary movement of the pharyngeal muscles allows food into the upper esophagus. Once inside the esophagus, the food is propelled towards the stomach via a peristaltic wave.

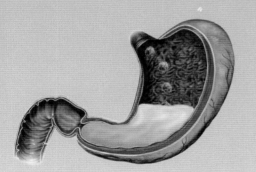 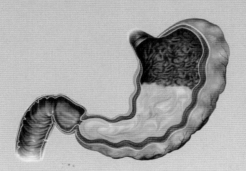 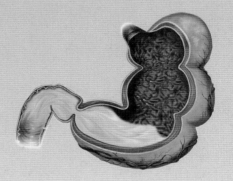

The stomach churns the food, combining it with gastric juices to form a semi-liquid known as chyme. The chyme is released periodically into the duodenum by the pyloric sphincter.

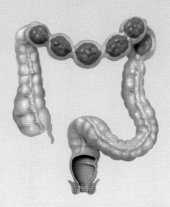 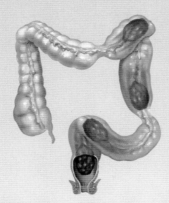 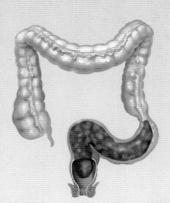

Nutrients are absorbed by the small intestines. In the colon, water and bile salts are absorbed before peristaltic contractions push the waste matter along to the rectum, where it is periodically expelled.

The ileum joins the large intestine at the ileocecal junction, with the large intestine consisting of the cecum, colon and rectum. The remains of the digested food travels along the large intestine, where water and bile salts are extracted, leaving only feces remaining.

The feces move on to the rectum and then to the anus, where they are periodically expelled.

The autonomic nervous system controls much of the digestive process, triggering the coordinated activity of muscles, organs, enzymes and hormones.

The Digestive Organs

THE ESOPHAGUS

This muscular tube transports food from the mouth to the stomach in wave-like contractions of the smooth muscle. It passes through the diaphragm to meet with the stomach, releasing food into the stomach via the lower esophageal sphincter.

THE STOMACH

The stomach has many muscular layers; these layers work together to churn food received from the esophagus, further breaking down the food into a semi-liquid form known as chyme. The mucosa and submucosa, which form the stomach's inner lining, secrete gastric juices to aid digestion. The layers of stomach muscle contract and expand in order to mix and expel the stomach contents, while the outer layer of the stomach is smooth, allowing ease of movement.

THE LIVER

The liver produces bile, which assists in the breakdown of partially digested food. The bile is sent from the liver to the gallbladder, where it is stored and concentrated ready for release into the duodenum.

THE PANCREAS

Pancreatic enzymes are responsible for much of the digestive process carried out in the duodenum. The pancreatic enzymes and bile enter the duodenum through a common opening, the duodenal ampulla.

THE INTESTINES

The small intestine consists of the duodenum, jejunum and ileum. The large intestine consists of the cecum, colon, rectum and anus. Most of the absorption of nutrients takes place in the small intestine, while the colon extracts water and bile salts. All that then remains is feces which is sent on to the rectum for periodic evacuation via the anus.

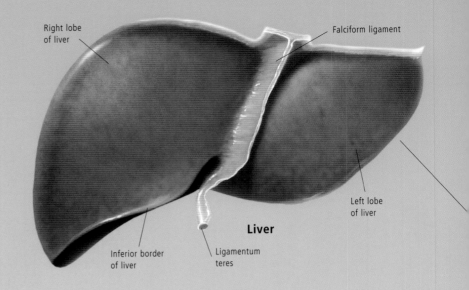

Right lobe of liver

Falciform ligament

Left lobe of liver

Inferior border of liver

Ligamentum teres

Liver

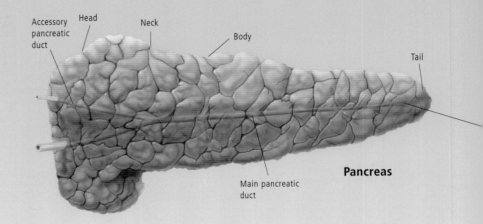

Accessory pancreatic duct

Head

Neck

Body

Tail

Main pancreatic duct

Pancreas

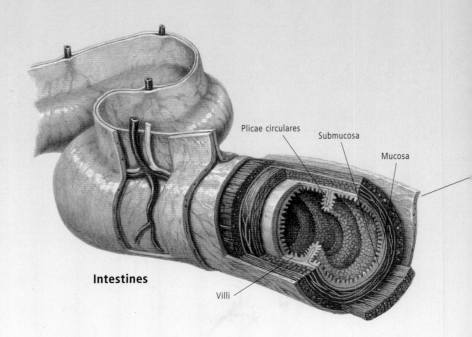

Plicae circulares

Submucosa

Mucosa

Intestines

Villi

Esophagus

Part of the digestive system, the esophagus is a muscular tube for transporting food from the pharynx to the stomach. The muscles of the esophagus are involuntary muscles controlled by the autonomic nervous system.

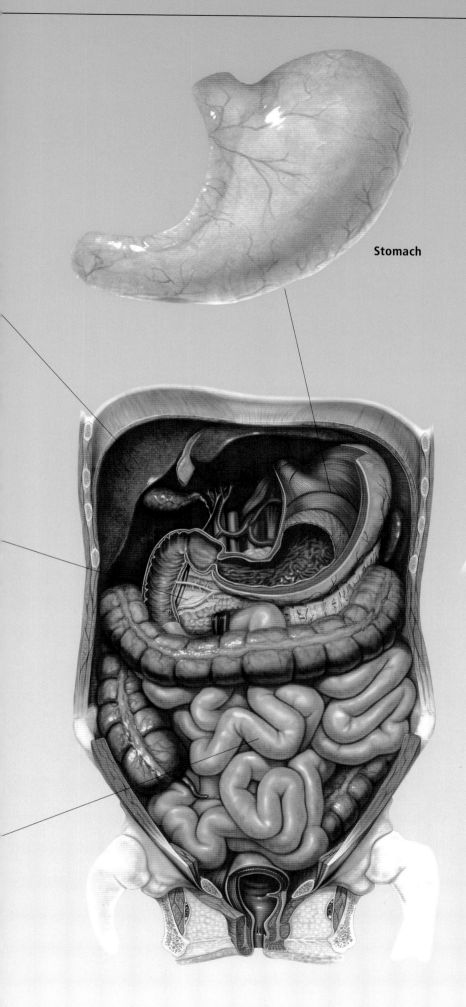

Stomach

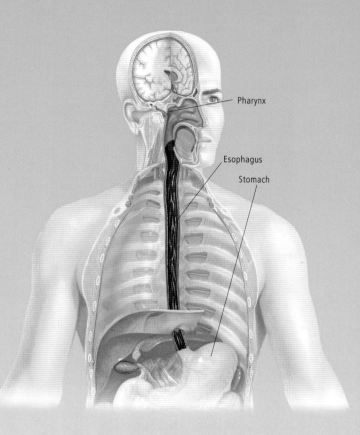

Pharynx

Esophagus

Stomach

DID YOU KNOW?

The stomach produces a thick layer of mucus to protect it from its own acid juices. The parietal or oxyntic cells of the inner lining of the stomach produce hydrochloric acid—this acid is strong enough to dissolve a nail.

Digestive organs

The digestive organs consist of the alimentary tract (a muscular tube that extends from the mouth to the anus) and the accessory organs (including the liver, gallbladder and pancreas).
NB: In this illustration the liver has been lifted up to show the gallbladder.

The Respiratory System

Our bodies require oxygen to build and regenerate cells. The body consumes oxygen and produces carbon dioxide as a waste product. Air inhaled from the outside environment is taken into the lungs, and the oxygen is exchanged for carbon dioxide, which is then exhaled. In order to complete this process, the respiratory system requires the coordinated efforts of the nose, pharynx, larynx, trachea and bronchi, in conjunction with the intercostal muscles and the diaphragm.

The upper part of the respiratory tract consists of the nose, the nasal cavity and the pharynx. To ensure particles do not enter the airways, the inner surface of the nostril is lined with hairs which capture inhaled particles.

The pharynx is a common passageway of both the respiratory and digestive systems. When food is swallowed, the epiglottis seals off the larynx to prevent food from entering the airways.

How we breathe

When we breathe in, air passes through the nose, pharynx, larynx, trachea and bronchi. As it does, the ribs rotate around their vertebral articulations, the sternum rises, and the diaphragm moves down—these muscular movements increase the capacity of the thoracic cavity, resulting in a decrease in intrathoracic pressure to a value below atmospheric pressure.

In the lungs, the bronchi subdivide into smaller and smaller bronchioles, which end with alveoli (air sacs). The sac walls are surrounded by small blood vessels which allow oxygen from the air to enter the bloodstream and carbon dioxide accumulated in the blood to pass into the alveoli to be breathed out.

When we breathe out—a passive process known as expiration—the ribs, lungs and diaphragm return to their original position, squeezing the air out.

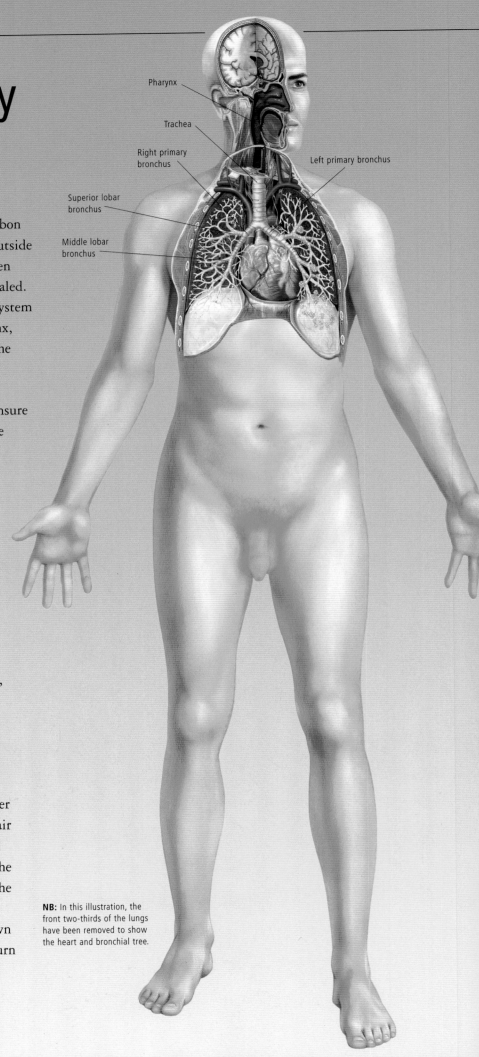

Pharynx

Trachea

Right primary bronchus

Left primary bronchus

Superior lobar bronchus

Middle lobar bronchus

NB: In this illustration, the front two-thirds of the lungs have been removed to show the heart and bronchial tree.

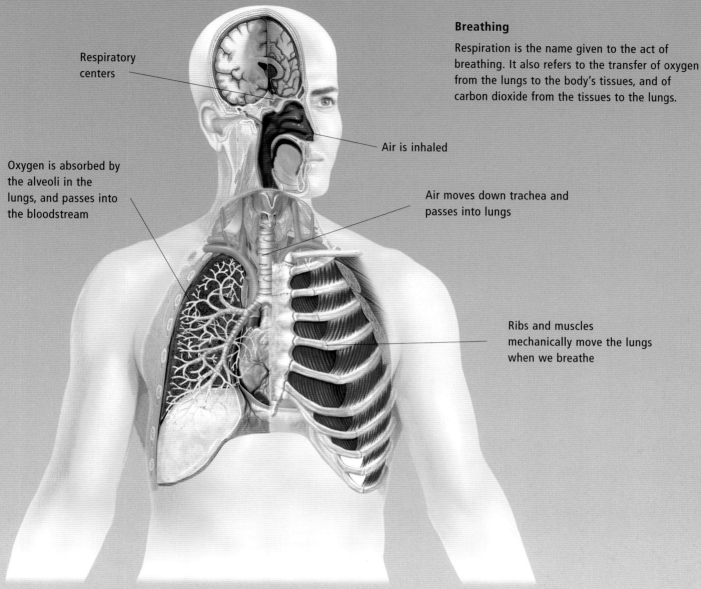

Respiratory centers

Breathing

Respiration is the name given to the act of breathing. It also refers to the transfer of oxygen from the lungs to the body's tissues, and of carbon dioxide from the tissues to the lungs.

Air is inhaled

Oxygen is absorbed by the alveoli in the lungs, and passes into the bloodstream

Air moves down trachea and passes into lungs

Ribs and muscles mechanically move the lungs when we breathe

DID YOU KNOW?

The lungs contain around 300 million air sacs (alveoli). The alveoli create a huge surface area, estimated to be 1,000 square feet (93 square meters), where the blood exchanges its carbon dioxide for a fresh supply of oxygen.

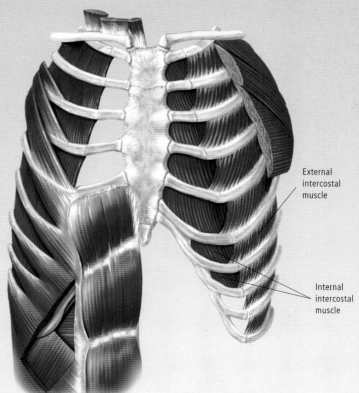

External intercostal muscle

Internal intercostal muscle

Intercostal muscles

The three layers of intercostal muscles contract to lift the sternum and expand the rib cage during inspiration. In the passive process of expiration, the ribs recoil back to their normal position.

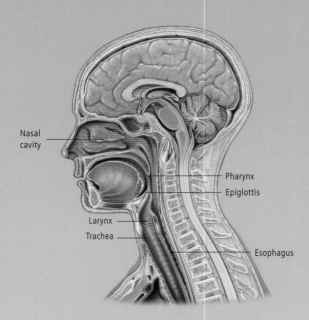

Nasal cavity

Pharynx

Epiglottis

Larynx

Trachea

Esophagus

Upper section of respiratory tract

The upper part of the respiratory tract includes the nose, nasal cavity, the pharynx and the larynx. The larynx leads to the trachea.

The Trachea, Bronchi and Lungs

The windpipe (trachea) is a $3^1/_2$–5 inch (9–12 centimeter) long tube comprising C-shaped cartilages interleaved with fibrous tissue. The trachealis muscle bridges the two ends of the cartilage to create a complete tube. The trachea branches into the two main bronchi, the left and right bronchi, leading to the left and right lungs.

The lungs are lobed, with the left lung having an upper and lower lobe, and the right lung having 3 lobes, the upper, middle and lower lobes. The bronchi enter the lungs and divide into the lobar bronchi, then subdivide into smaller and smaller bronchioles which terminate at the air sacs (alveoli). Bunched around the bronchioles, the thin-walled alveoli are separated from one another by interalveolar septa. This network of bronchi and bronchioles is known as the bronchial tree.

Gas exchange in the lungs takes place when oxygen from the inspired air and carbon dioxide found in blood from the pulmonary artery are exchanged. The gases move across the alveolar membrane—oxygen enters the blood via the capillary network of the alveoli, and carbon dioxide enters the alveoli, and is expelled during expiration.

Alveoli

The lungs contain over 300 million tiny air sacs (alveoli), providing a huge surface area for gas exchange.

Capillary network around alveoli

Branch of pulmonary vein

Visceral pleura

Endothoracic fascia

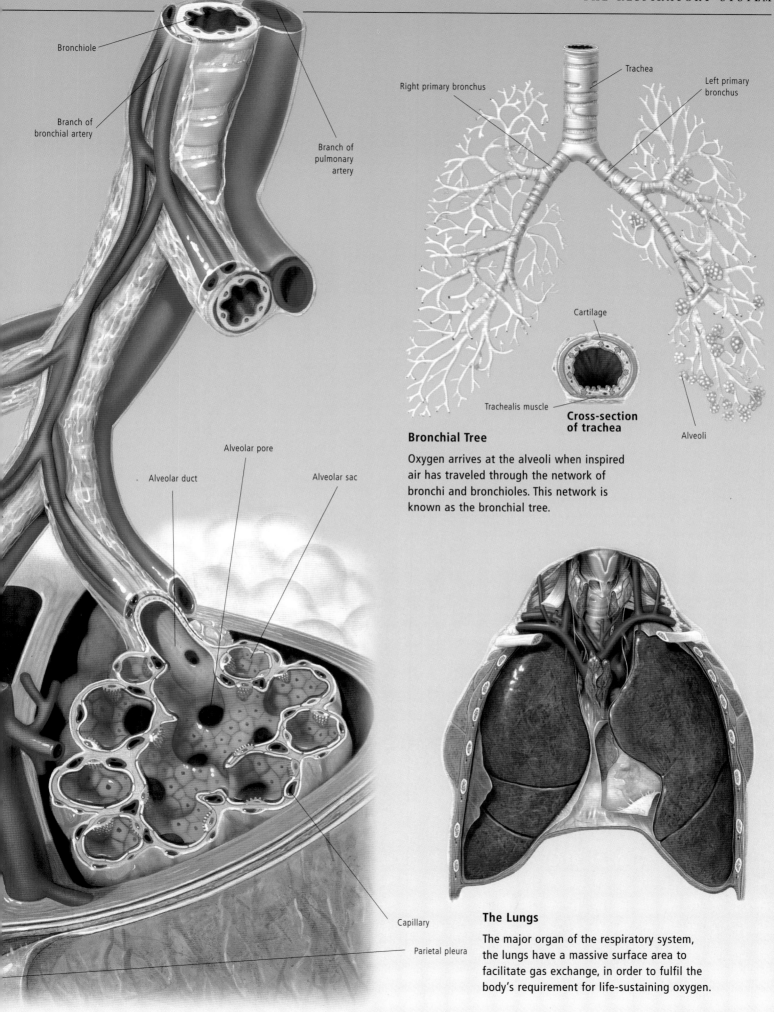

Bronchiole

Branch of bronchial artery

Branch of pulmonary artery

Alveolar pore

Alveolar duct

Alveolar sac

Capillary

Parietal pleura

Right primary bronchus

Trachea

Left primary bronchus

Cartilage

Trachealis muscle

Cross-section of trachea

Alveoli

Bronchial Tree

Oxygen arrives at the alveoli when inspired air has traveled through the network of bronchi and bronchioles. This network is known as the bronchial tree.

The Lungs

The major organ of the respiratory system, the lungs have a massive surface area to facilitate gas exchange, in order to fulfil the body's requirement for life-sustaining oxygen.

The Urinary System

The urinary system maintains the balance of water and electrolytes in the blood and excretes the waste products of metabolism.

The Kidneys

Located on the back wall of the abdomen, the kidneys lie either side of the vertebrae, with the right kidney slightly lower than the left. The adrenal glands sit on top of each of the kidneys. Each kidney has an outer cortex and an inner medulla. Within the cortex are the filtration units (nephrons) of the kidney, while the medulla contains about a dozen renal pyramids. The renal pyramids contain collecting tubules which collect the urine produced by the filtration units in the cortex.

The excretory function of the kidneys is to remove metabolic waste products, such as urea and any other unwanted substances. The balance of electrolytes in the blood and the water content of the body are also regulated by the kidneys. The kidney nephrons process filtrate (fluid composed of blood minus red and white blood cells and proteins), reabsorbing water, glucose and salts. Waste products are secreted into the filtrate, which is excreted from the body as urine. On average, the daily adult urine output is around 1–1 ½ quarts (1–1.5 liters).

The kidney also has an endocrine function, releasing hormones involved in blood formation and calcium metabolism.

The two large renal arteries supply blood and oxygen to the kidneys, with branches of the arteries supplying the adrenal gland and ureters. The renal arteries then branch again, becoming the anterior and posterior divisions of the arteries.

Each branch subdivides into smaller and smaller branches, eventually forming the capillaries which supply oxygen to the kidney tissue and participate in kidney filtration.

Male urinary system

The male urinary system comprises the kidneys, the ureters, the bladder and the urethra. The urethra is the common passageway for both sperm and urine.

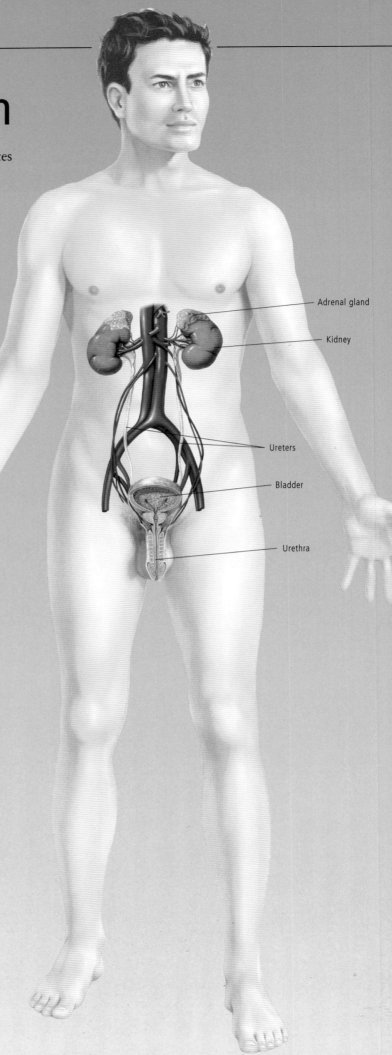

Adrenal gland

Kidney

Ureters

Bladder

Urethra

The Ureter

Continuous with the renal pelvis of the kidney, the ureter is a muscular tube transporting urine to the bladder. The ureters enter the bladder near the midline; they run obliquely for about ¾ inch (2 centimeters) in the bladder before opening to slit-like apertures. To prevent backflow of urine, the full bladder compresses the part of the ureter in the bladder wall.

The Male Urethra

With a role in both the urinary and reproductive systems, the male urethra is a muscular tube that provides a passageway for both urine and for sperm from the reproductive system. From the bladder, it passes through the prostate and continues on to the penis.

Kidneys

The kidneys filter the blood, absorbing water and electrolytes for return to the circulatory system, and transferring waste products (urine) to the bladder via the ureters. Usually, the right kidney is slightly lower than the left kidney. The tip of each kidney is capped by the adrenal glands.

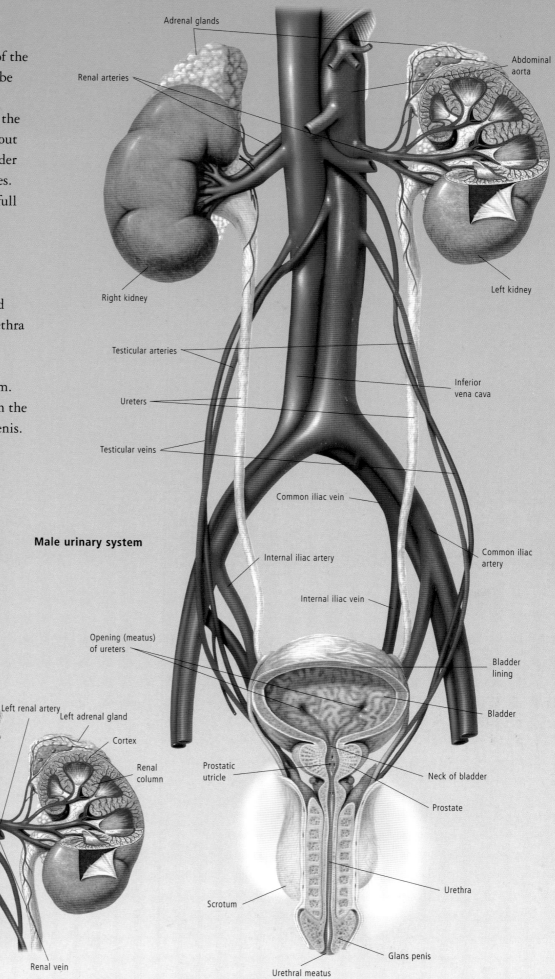

Adrenal glands

Renal arteries

Abdominal aorta

Right kidney

Left kidney

Testicular arteries

Inferior vena cava

Ureters

Testicular veins

Common iliac vein

Common iliac artery

Male urinary system

Internal iliac artery

Internal iliac vein

Opening (meatus) of ureters

Bladder lining

Bladder

Prostatic utricle

Neck of bladder

Prostate

Urethra

Scrotum

Glans penis

Urethral meatus

Right adrenal gland

Renal pyramid (medulla)

Left renal artery

Left adrenal gland

Cortex

Renal papilla

Renal column

Renal cortex

Major calyx

Minor calyx

Renal pelvis

Renal vein

The Bladder

Urine produced in the kidneys is transferred to the bladder via the ureters. The bladder is a muscular sac, with a capacity of about 1 pint (475 milliliters). The male and female bladders are similar in structure, but the male bladder sits on top of the prostate.

The Female Urinary System

While the female urinary system is essentially the same as in the male, there are a few differences. The female bladder lies in the pelvic cavity, with the uterus lying on top of the bladder. A relatively short urethra joins the bladder to the external environment, with the opening lying in front of the entrance to the vagina. Unlike the male system, the female urethra plays no part in the reproductive system.

Kidneys in situ

The kidneys lie on the back wall of the abdomen on either side of the vertebrae. The right kidney lies behind the duodenum and is slightly lower than the left, which lies behind the head of the pancreas and the stomach. In front of the kidneys are the ends of the transverse colon, where it joins the ascending and descending colon.

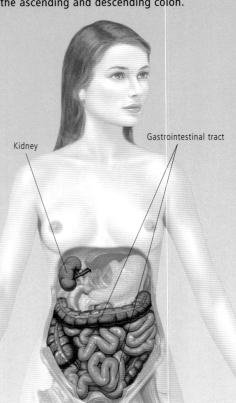

Kidney

Gastrointestinal tract

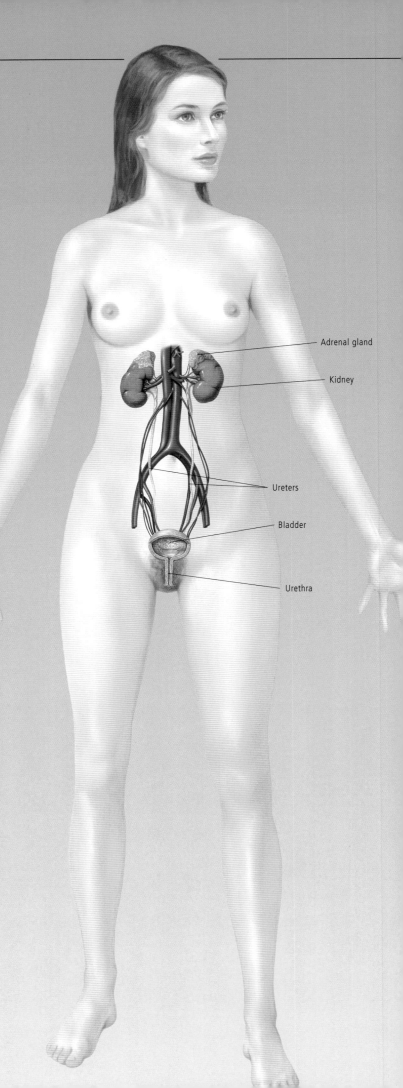

Adrenal gland

Kidney

Ureters

Bladder

Urethra

DID YOU KNOW?

Each day, the kidneys process and filter gallons of blood through more than 2 million nephrons. The filtering ability of the nephrons enables around 99 percent of blood to be recycled by the body.

Female urinary system

The female urinary system is similar in most ways to that of the male. However the urethra is much shorter, passing directly through the pelvic floor to the outside environment via an opening in front of the vagina. Unlike the male urinary system, the female urethra is not involved in the reproductive system.

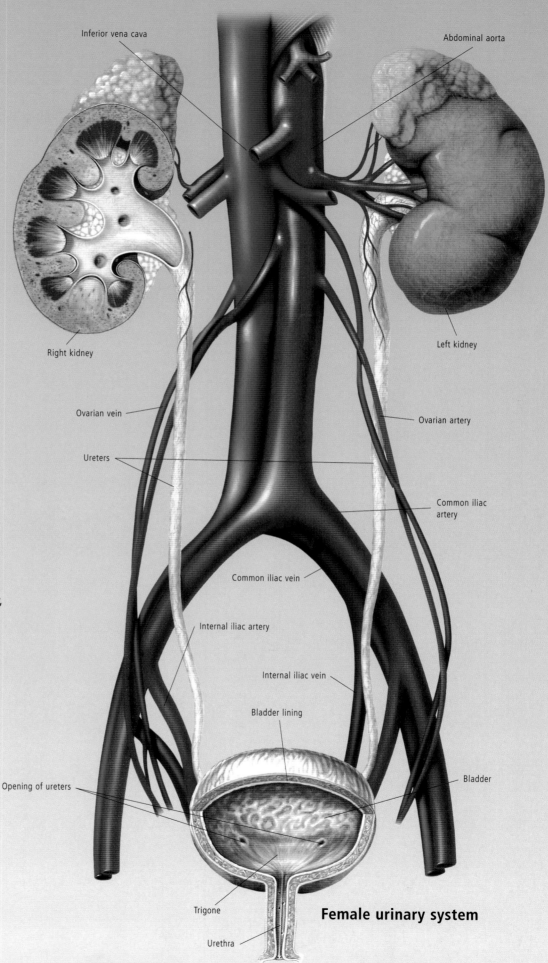

Inferior vena cava

Abdominal aorta

Right kidney

Left kidney

Ovarian vein

Ovarian artery

Ureters

Common iliac artery

Common iliac vein

Internal iliac artery

Internal iliac vein

Bladder lining

Bladder

Opening of ureters

Trigone

Urethra

Female urinary system

The Male Reproductive System

The male reproductive system includes the testes, the ductus deferens, the seminal vesicle, the prostate gland and the penis.

Located inside the scrotum, the sac that lies just behind the penis, the testes are the major organs of reproduction. They manufacture the male sex hormones, testosterone and androsterone, and produce sperm. During the development of the fetus (at about 30 weeks) the testes move from their original position near the kidneys, down the inguinal canal, to the scrotum. The scrotum is cooler than the rest of the body: sperm do not properly develop at normal body temperature. The epididymis is an oblong structure attached to the upper part of each testis.

The ductus deferens (vas deferens) is the tube that can be felt above the testis through the loose part of the scrotum.

The seminal vesicle is a single coiled tube that lies outside the ductus deferens. The duct of the seminal vesicle fuses with the ductus deferens to form the ejaculatory duct. The seminal vesicle manufactures more than half of the semen.

Shaped like an inverted pyramid and about the size of a walnut, the prostate surrounds the neck of the bladder and the urethra. It is composed of muscle and glandular tissue. Secretions from both the seminal vesicle and the prostate constitute the seminal fluid ejaculated during orgasm. Seminal fluid contains the glucose and enzymes that provide the energy needed for the sperm to reach the ovum.

The penis is composed of two cylinders of sponge-like vascular tissue (corpora cavernosa). A third cylinder (corpus spongiosum) contains the urethra, which ends in the glans, the bulbous external swelling at the tip of the penis. In an

Reproductive system: male

The male reproductive system is comprised of the testes, the ductus deferens, the seminal vesicle, the prostate gland and the penis.

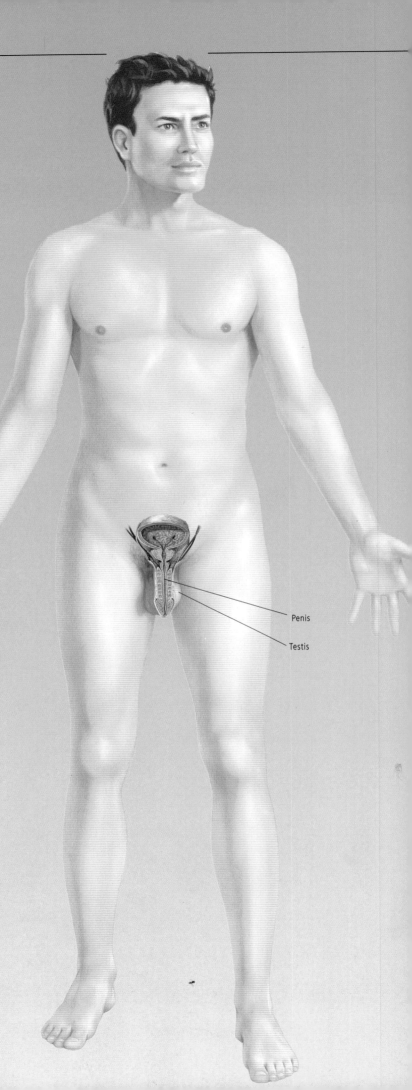

Penis

Testis

uncircumcised penis, the glans is covered by a protective foreskin (prepuce). All three cylinders are encased in thick connective tissue and are structured like a sponge, the interconnecting spaces containing blood. After sexual stimulation, the two spongy cylinders become engorged with blood, and the penis becomes hard and erect. This enables the man to insert his erect penis into a woman's vagina. When excitement during intercourse reaches a peak, the sympathetic nervous system triggers an emission of semen (ejaculation) which contains several million sperm. There must be an adequate number of healthy, active sperm in the semen for fertilization to occur.

Sperm are microscopic cells that carry the male genetic material to join with that of the female after fertilization of her ovum. During ejaculation, the sperm combine with secretions from the prostate and seminal vesicles to form the seminal fluid. The head of each sperm has a nucleus containing chromosomes, and an acrosomal membrane that holds enzymes required for fertilization. The tail of the sperm helps it move from the testes to the female reproductive organs.

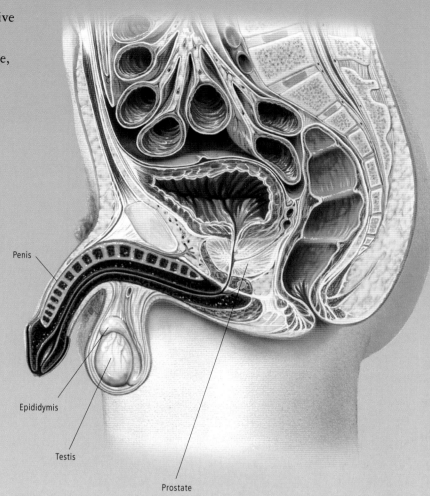

Penis

Epididymis

Testis

Prostate

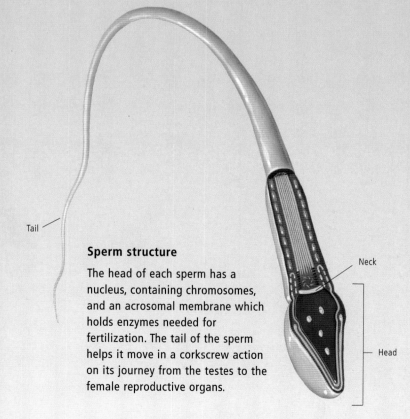

Sperm structure

The head of each sperm has a nucleus, containing chromosomes, and an acrosomal membrane which holds enzymes needed for fertilization. The tail of the sperm helps it move in a corkscrew action on its journey from the testes to the female reproductive organs.

Tail

Neck

Head

Testes

The testes lie directly behind the penis, in the scrotum. They are the male gonads, and produce the male sex hormones and sperm. Sperm are produced in the tubules in the testes. Sperm cells divide and produce spermatids, which mature into spermatozoa. Sperm move through the testes and into the epididymis; they stay here until they are mature and ready for ejaculation.

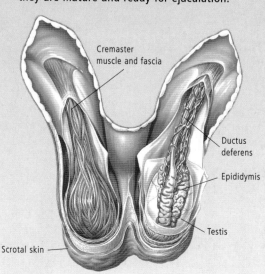

Cremaster muscle and fascia

Ductus deferens

Epididymis

Testis

Scrotal skin

The Female Reproductive System

The female reproductive system consists of the ovaries, the fallopian tubes, the uterus and the vagina.

The ovaries are almond-shaped organs approximately 1½ inches (3 centimeters) long and ½ inch (1 centimeter) wide. They are located on either side of the uterus and are supported by the broad ligament. Each ovary contains thousands of undeveloped follicles, each containing an ovum. Ovulation takes place in the middle of a menstrual cycle: a Graafian follicle ruptures, releasing an ovum, which enters the fallopian tubes.

Each of the fallopian tubes leads from the ovary. The tubes are trumpet-shaped, with the narrow end joined to the uterus and the wider flared end next to the ovary. The fallopian tubes transport the ovum to the uterus. If a sperm penetrates the ovum, fertilization takes place in the outer third of the fallopian tube. When the ovum is fertilized it will implant itself into the lining of the uterus and develop into a baby.

The uterus (womb) is situated between the bladder and the rectum. Normally (in a non-pregnant woman) the uterus is pear-shaped and flattened from front to back. The upper two-thirds is the body; the lower third is the cervix, a muscular tube that extends into the vagina. The endometrium is the inner lining of the uterus, and is shed at the end of the menstrual cycle. If the egg is fertilized, the endometrium prepares for the implantation of the fertilized ovum. Part of the endometrium becomes the placenta, which nourishes and protects the unborn baby. In pregnancy, the uterus expands to accommodate the developing fetus.

The vagina is a fibromuscular tube that connects the uterus to the outside of the body. The clitoris is a small protrusion that is very sensitive to sexual stimulation; the prepuce of the clitoris is an extension of the labia minora, which lie on each side of the vagina.

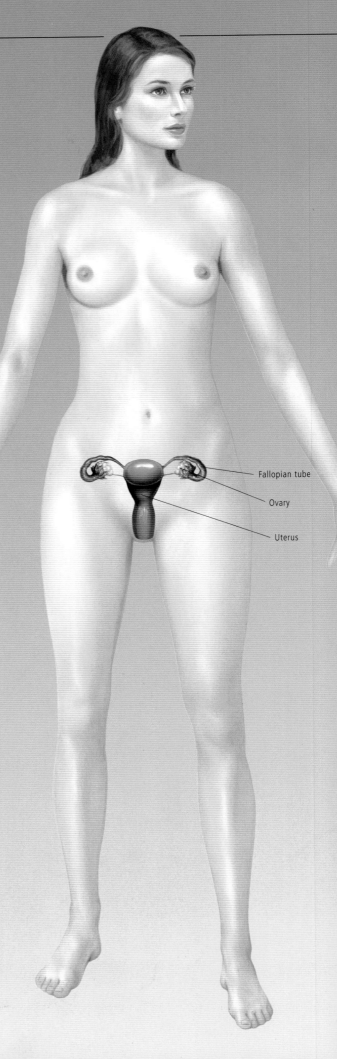

Fallopian tube

Ovary

Uterus

Reproductive system: female

The female reproductive system is composed of two ovaries that produce eggs (ova) and female hormones, two fallopian tubes (uterine tubes), the uterus, and the vagina, which extends from the cervix to the vulva.

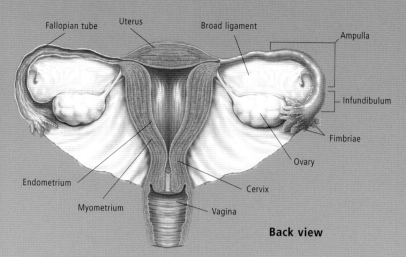

Back view

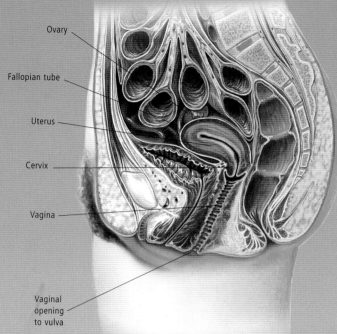

Uterus

The uterus resembles a slightly flattened, upside-down pear, with a slight constriction dividing it in two. Situated in the pelvis between the bladder and the rectum, the uterus is the site where the fertilized egg grows into an embryo and fetus. The upper two-thirds of the uterus is the body and the lower third is the cervix.

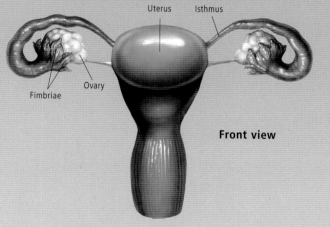

Front view

Female reproductive organs

Ovaries

The ovaries are the female sex organs in which the eggs (ova) are formed. They produce estrogen and progesterone. The ovaries are each about the shape and size of an almond, and are located on either side of the uterus. In the middle of a menstrual cycle, the ovary produces an ovum, which enters the fallopian tube and moves towards the uterus. If fertilization occurs, it takes place in the outer third of the tube; the fertilized ovum implants itself into the lining of the uterus.

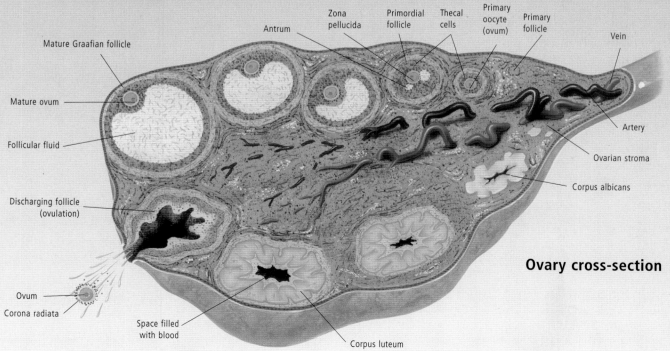

Ovary cross-section

The Endocrine System

The endocrine system consists of the pineal body, thymus, thyroid, parathyroids, adrenals, pancreas, ovaries and testes, all under the control of the pituitary gland. The endocrine system releases hormones which control the activities of body tissues. Hormones, made of amino acids or steroids, are released from endocrine cells at specific times and in measured amounts to act on target organs; the changes brought about by the release of these hormones are often slow or long-term changes. Hormones are released directly into the bloodstream or body cavities, and each hormone acts on specific target areas, often at some distance from the source.

The relationship between the chief organ of the endocrine system, the pituitary gland, and the hypothalamus in the brain means that the nervous system and endocrine system are involved in control of body functions. Endocrine hormones affect the nervous system and many endocrine organs are stimulated or inhibited by nerve cells.

Responsible for coordinating the activities of the system, the pituitary gland is divided into two lobes, an anterior lobe and a posterior lobe. The anterior lobe is responsible for the production of growth hormones, prolactin, follicle-stimulating hormone, luteinizing hormone, thyroid-stimulating hormone, adrenocorticotrophic hormone and melanocyte-stimulating hormone. Particularly important during childhood and early adolescence, growth hormones stimulate the growth of long bones. Prolactin stimulates the mammary glands of the breast, to promote and maintain milk production. Follicle-stimulating hormone (FSH) stimulates the production of eggs in women and sperm in men, while

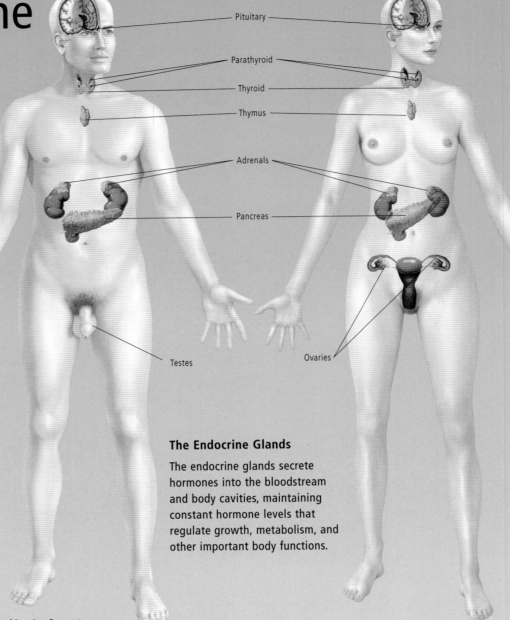

Pituitary

Parathyroid

Thyroid

Thymus

Adrenals

Pancreas

Testes

Ovaries

The Endocrine Glands

The endocrine glands secrete hormones into the bloodstream and body cavities, maintaining constant hormone levels that regulate growth, metabolism, and other important body functions.

luteinizing hormone stimulates the release of the eggs and the production of progesterone in women, and the secretion of testosterone in men. The thyroid is triggered into hormone production by thyroid-stimulating hormone (TSH) in the pituitary gland, while the adrenal gland is triggered into action by adrenocorticotrophic hormone.

The posterior lobe of the pituitary contains oxytocin and antidiuretic hormones, produced in the hypothalamus and transported to the pituitary within nerve fibers. Oxytocin encourages the contraction of the smooth muscle cells in the uterus and around the milk glands in the breasts. Antidiuretic hormone (vasopressin) promotes the reabsorption of water from the urine in the kidney, thus controlling the salt levels in the blood.

Hormones produced by the thyroid gland act to increase energy production; they also affect the developing brain. Parafollicular cells or C cells in the thyroid produce calcitonin, responsible for reducing the concentration of calcium in the blood. The hormones produced by the parathyroid gland work to raise the concentration of calcium in the blood, and also reduce the concentration of phosphate ions.

The adrenal glands have two layers, a cortex and a medulla. Hormones produced in the medulla are released in response to threatening or intense emotional situations, raising the blood sugar level, elevating blood pressure and quickening the heartbeat. The adrenal cortex produces three main types of hormones: glucocorticoids, which promote the breakdown of protein and the release of fat and sugars into the blood stream; mineralocorticoids, which promote the absorption of sodium in the kidney; and sex steroids.

Areas within the pancreas, known as the pancreatic islets, or islets of Langerhans, produce hormones responsible for controlling blood sugar levels; insulin acts to lower blood glucose concentration, while glucagon acts to raise it.

The tiny pineal body, found in the skull cavity, surrounded by the brain, produces melatonin. The concentrations of melatonin vary around the 24-hour cycle of the day (circadian rhythm).

Estrogen and progesterone are produced by the ovaries, and undergo changes in level during a 28-day cycle. Estrogen promotes growth of the breasts and reproductive organs, while progesterone maintains the lining of the uterus in readiness for implantation by a fertilized ovum. Estrogen and progesterone are triggered by FSH and luteinizing hormone from the pituitary. During pregnancy the placenta acts as an endocrine organ, producing hormones to sustain the pregnancy and promote fetal growth.

The body has various mechanisms in place to regulate and monitor hormone production, thus ensuring correct levels are maintained.

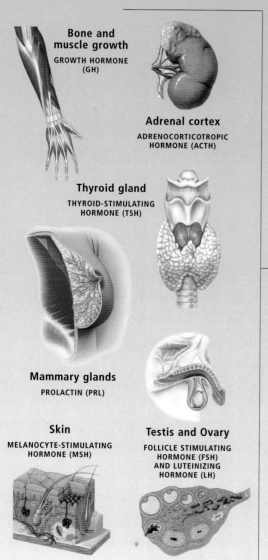

Bone and muscle growth
GROWTH HORMONE (GH)

Adrenal cortex
ADRENOCORTICOTROPIC HORMONE (ACTH)

Thyroid gland
THYROID-STIMULATING HORMONE (TSH)

Mammary glands
PROLACTIN (PRL)

Skin
MELANOCYTE-STIMULATING HORMONE (MSH)

Testis and Ovary
FOLLICLE STIMULATING HORMONE (FSH) AND LUTEINIZING HORMONE (LH)

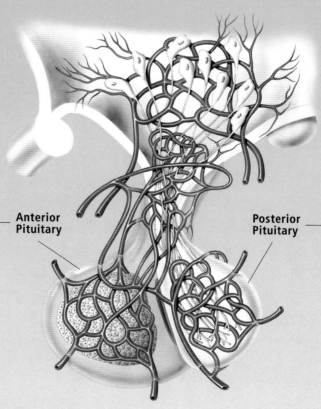

Anterior Pituitary

Posterior Pituitary

The pituitary gland

The pituitary gland is the center of control for the endocrine system, regulating the function and operation of the other endocrine organs. The illustration shows the organs and tissues controlled by the pituitary gland.

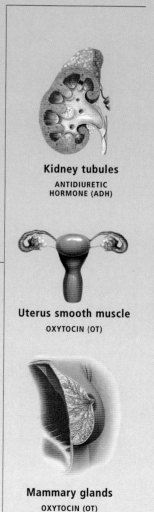

Kidney tubules
ANTIDIURETIC HORMONE (ADH)

Uterus smooth muscle
OXYTOCIN (OT)

Mammary glands
OXYTOCIN (OT)

The Skin

The skin is a protective organ that covers the body, and is the body's largest organ. The function of the skin is to provide a protective layer against injury and attack, extremes of temperature, and invading organisms such as viruses and bacteria. The skin also factors in temperature regulation, vitamin D production and ultraviolet protection.

Within its three layers, the epidermis, dermis and subcutaneous tissue, the skin contains specialized structures, including nerve receptors, hair follicles, sweat glands and sebaceous glands.

The outer layer of the skin, the epidermis, is itself divided into 5 layers, with each of the sublayers performing

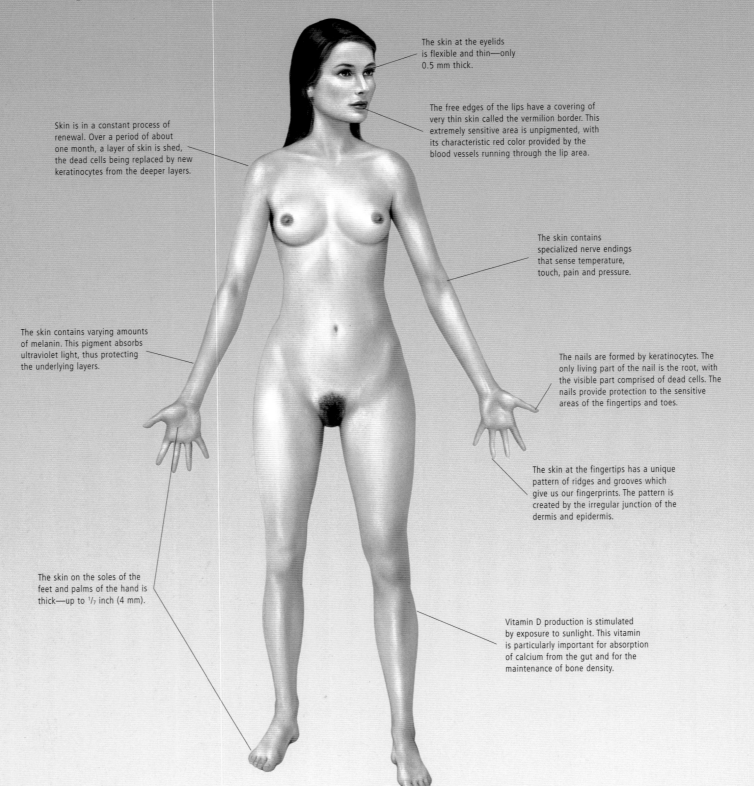

The skin at the eyelids is flexible and thin—only 0.5 mm thick.

The free edges of the lips have a covering of very thin skin called the vermilion border. This extremely sensitive area is unpigmented, with its characteristic red color provided by the blood vessels running through the lip area.

Skin is in a constant process of renewal. Over a period of about one month, a layer of skin is shed, the dead cells being replaced by new keratinocytes from the deeper layers.

The skin contains specialized nerve endings that sense temperature, touch, pain and pressure.

The skin contains varying amounts of melanin. This pigment absorbs ultraviolet light, thus protecting the underlying layers.

The nails are formed by keratinocytes. The only living part of the nail is the root, with the visible part comprised of dead cells. The nails provide protection to the sensitive areas of the fingertips and toes.

The skin at the fingertips has a unique pattern of ridges and grooves which give us our fingerprints. The pattern is created by the irregular junction of the dermis and epidermis.

The skin on the soles of the feet and palms of the hand is thick—up to $\frac{1}{7}$ inch (4 mm).

Vitamin D production is stimulated by exposure to sunlight. This vitamin is particularly important for absorption of calcium from the gut and for the maintenance of bone density.

a specific function. The bottom layer (stratum basale) is responsible for the production of melanin, which absorbs dangerous ultraviolet light and creates the pigment which gives skin its tanned appearance after exposure to sunlight. The outermost layer (stratum corneum), provides the main defence against skin infection. This outer layer consists mainly of dead cells, which are constantly sloughed off, being replaced by the cells of the underlying layer. This process means that the skin is constantly being renewed.

The structures lying within the layers of the skin each perform a supportive role for the skin. The hair follicles hold the hairs that offer protection to the skin and form an insulating layer over the skin in cold conditions. Usually allied to the hair follicles are the sebaceous glands. These glands release sebum, a liquid which lubricates and softens the skin. The sweat glands secrete a watery fluid onto the surface of the skin. Found almost everywhere on the body, the small sweat glands open at pores onto the surface of the skin. When the body becomes overheated, the sweat glands are activated, releasing sweat to cool the skin.

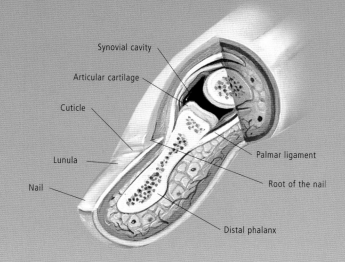

Nail

The function of the nails is to protect the sensitive tips of our fingers and toes. Comprised mostly of dead cells, the only living part is the root, which lies under a flap of skin, the cuticle. The nail bed lies beneath the nail, which acquires its characteristic pink color from the blood vessels running through the fingertip.

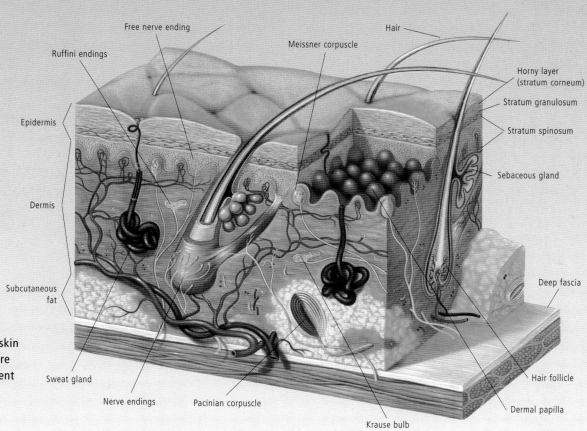

Skin

Near the surface of the skin (the horny layer), cells are flattened. The arrangement of cell layers provides a protective shield and prevents dehydration.

The Body Regions

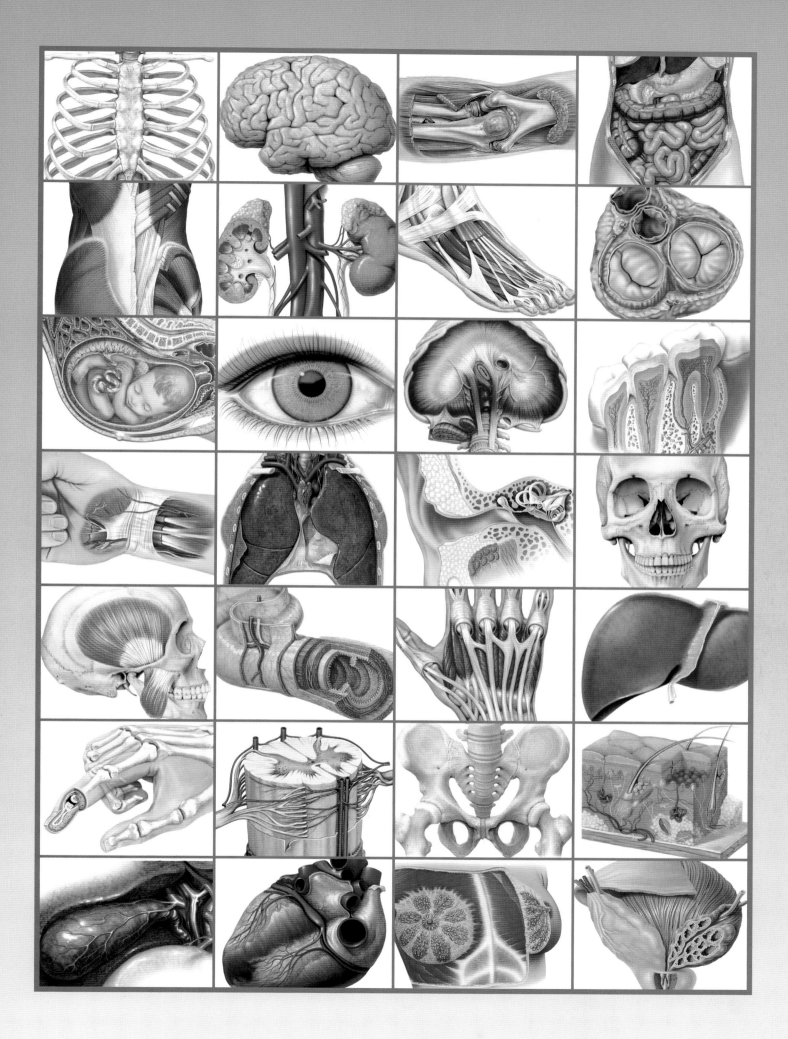

The Head

The head contains the brain, within the protective shell of the skull. Strictly speaking, the skull comprises two parts, the cranium and the face, but the term skull is often used to refer to the cranium, which covers the brain.

Controlled by the cranial nerves in the brain, the special sense organs of the nose, eyes, mouth and ears provide the senses of smell, sight, taste, hearing and balance.

The mouth is also involved in both the digestive system and the respiratory system, allowing the passage of air, food and liquids.

The cervical vertebrae of the spine, combined with the muscles of the neck, support the head and permit great freedom of movement.

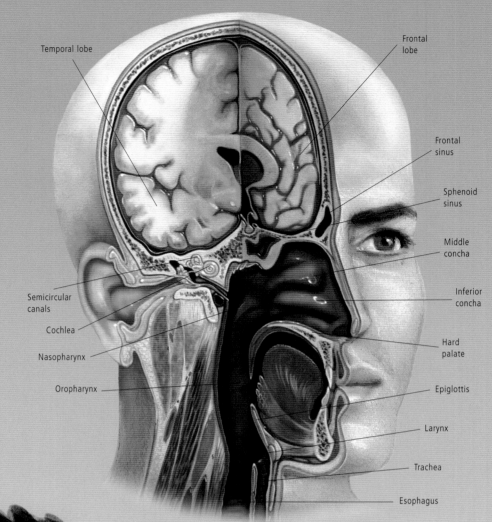

Temporal lobe
Frontal lobe
Frontal sinus
Sphenoid sinus
Middle concha
Inferior concha
Hard palate
Epiglottis
Larynx
Trachea
Esophagus
Semicircular canals
Cochlea
Nasopharynx
Oropharynx

Head

The head contains the brain and the special sense organs of sight, smell, hearing, balance and taste. Muscles in the neck enable the head to flex, extend and partially rotate.

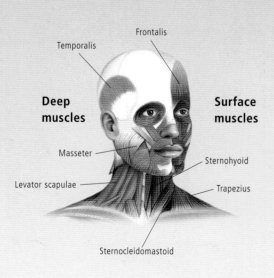

Frontalis
Temporalis
Deep muscles
Surface muscles
Masseter
Sternohyoid
Levator scapulae
Trapezius
Sternocleidomastoid

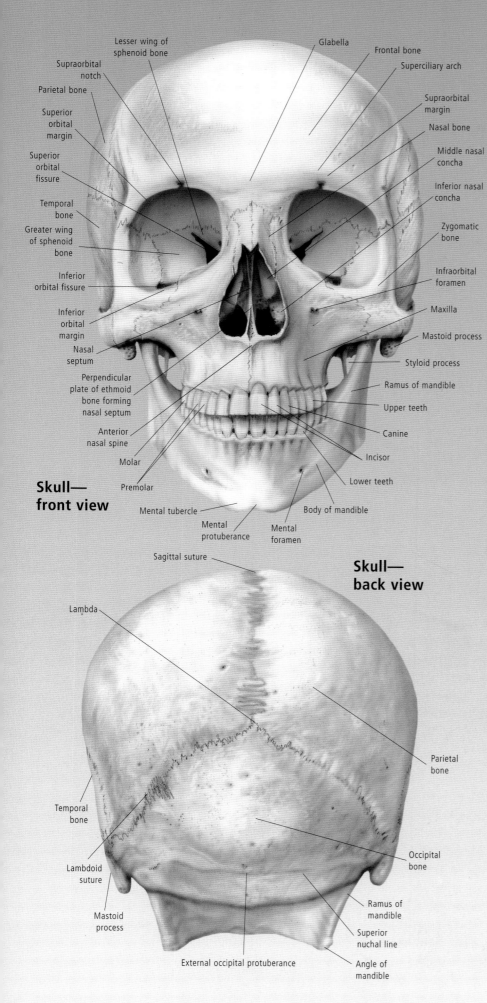

Skull—front view

Lesser wing of sphenoid bone
Supraorbital notch
Parietal bone
Superior orbital margin
Superior orbital fissure
Temporal bone
Greater wing of sphenoid bone
Inferior orbital fissure
Inferior orbital margin
Nasal septum
Perpendicular plate of ethmoid bone forming nasal septum
Anterior nasal spine
Molar
Premolar
Mental tubercle
Mental protuberance
Mental foramen
Glabella
Frontal bone
Superciliary arch
Supraorbital margin
Nasal bone
Middle nasal concha
Inferior nasal concha
Zygomatic bone
Infraorbital foramen
Maxilla
Mastoid process
Styloid process
Ramus of mandible
Upper teeth
Canine
Incisor
Lower teeth
Body of mandible

Skull—back view

Sagittal suture
Lambda
Temporal bone
Lambdoid suture
Mastoid process
External occipital protuberance
Parietal bone
Occipital bone
Ramus of mandible
Superior nuchal line
Angle of mandible

The Skull

The skull forms the skeleton of the head and is part of the axial skeleton. The most complex bony structure in the body, the skull serves to protect the brain, eyes and inner ears; forms the upper and lower jaws; and provides attachment for muscles of the face, eyes, tongue, pharynx and neck.

With the exception of the lower jaw, the bones of the skull are joined to each other by joints called sutures. The edges of the bones lock together, rather like a jigsaw puzzle, and are secured in place by fibrous connective tissue.

The bones of the cranium, housing the brain, are the frontal bone, the paired parietal and temporal bones and the occipital bone, with the sphenoid bone also forming part of the cranial joint. These bones (with the exception of the parietal bones) form the floor of the cranium. A system of spaces within the temporal bone forms the middle and inner parts of the ear. Separating the skull and the brain are three layers of membrane, known as the meninges.

The bones of the front of the skull that form the face include the frontal bone, the zygomatic bone, the maxillae and the mandible.

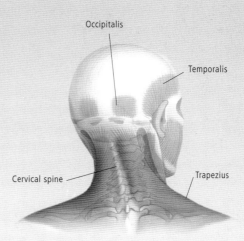

Occipitalis
Temporalis
Cervical spine
Trapezius

Posterior view—muscles of the head and neck

Skull— side view

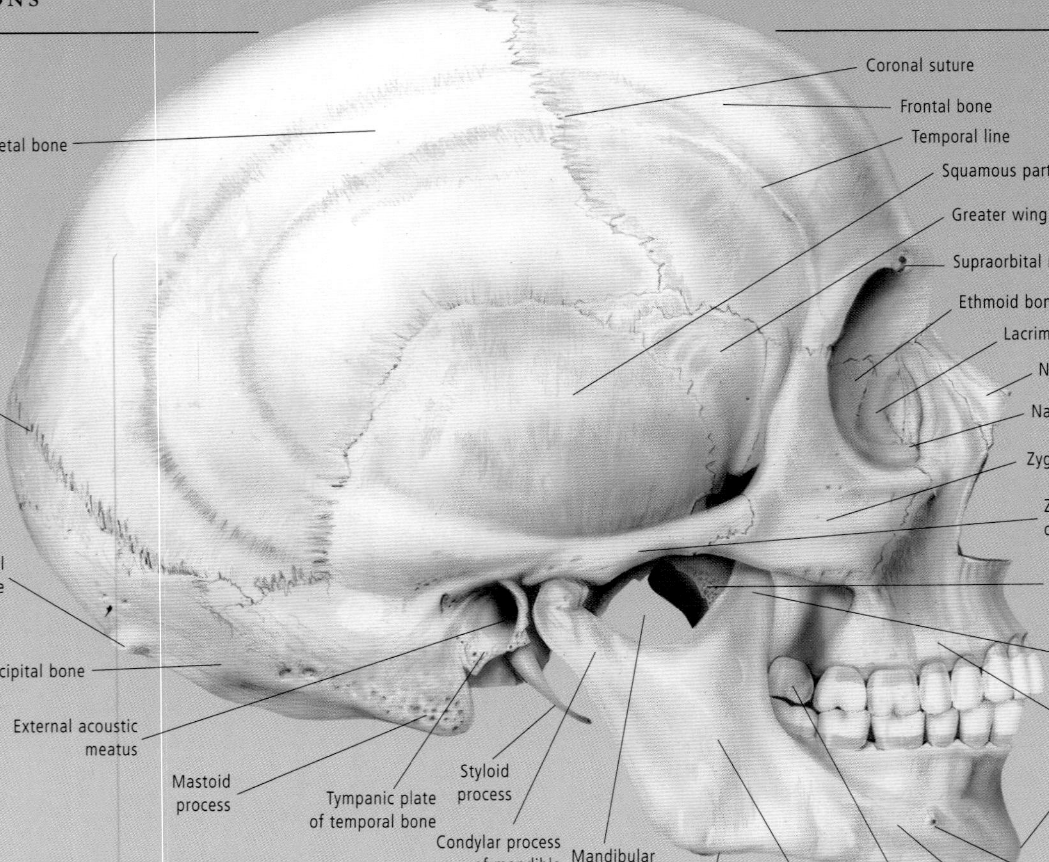

Parietal bone

Coronal suture

Frontal bone

Temporal line

Squamous part of temporal bone

Greater wing of sphenoid bone

Supraorbital notch

Ethmoid bone

Lacrimal bone

Nasal bone

Nasolacrimal duct

Zygomatic bone

Zygomatic process of temporal bone

Pterygoid process of sphenoid bone

Coronoid process of mandible

Maxilla

Lambdoid suture

External occipital protuberance

Occipital bone

External acoustic meatus

Mastoid process

Tympanic plate of temporal bone

Styloid process

Condylar process of mandible

Mandibular notch

Angle of mandible

Ramus of mandible

Wisdom tooth

Body of mandible

Mental protuberance

Mental foramen

Skull—top

Frontal bone

Coronal suture

Sagittal suture

Occipital bone

Parietal bone

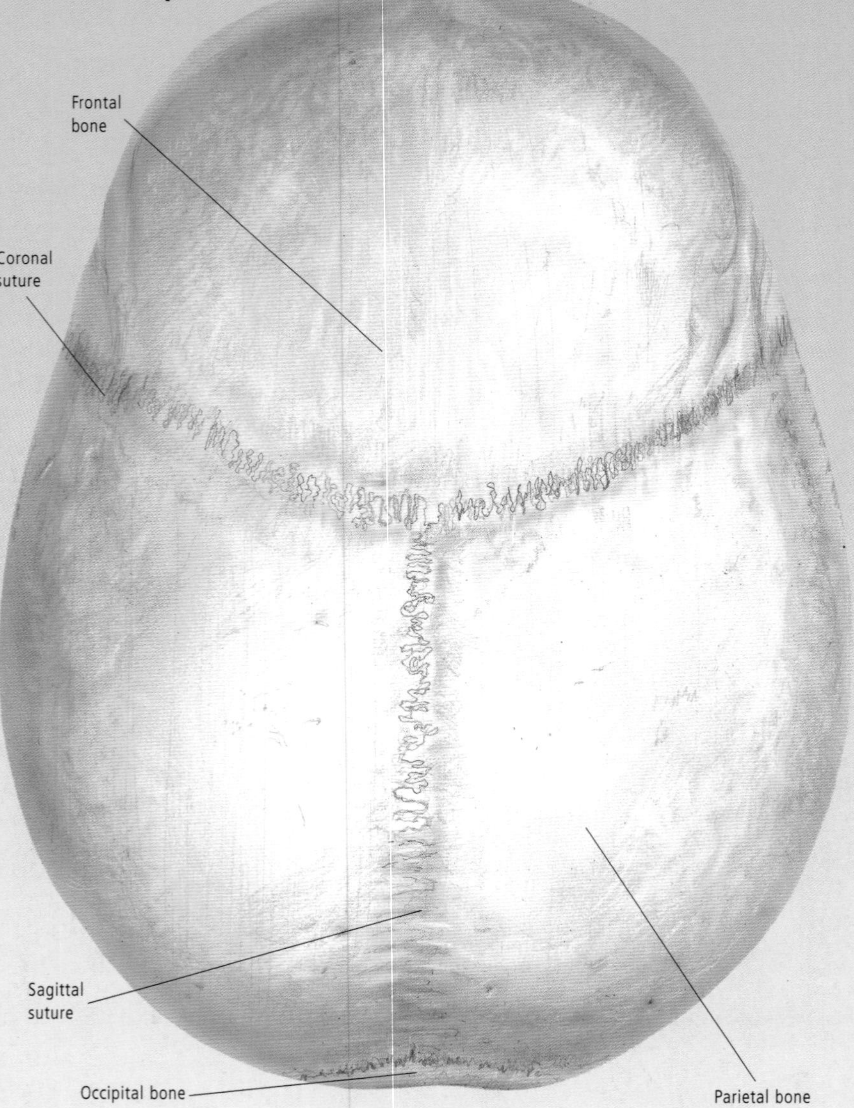

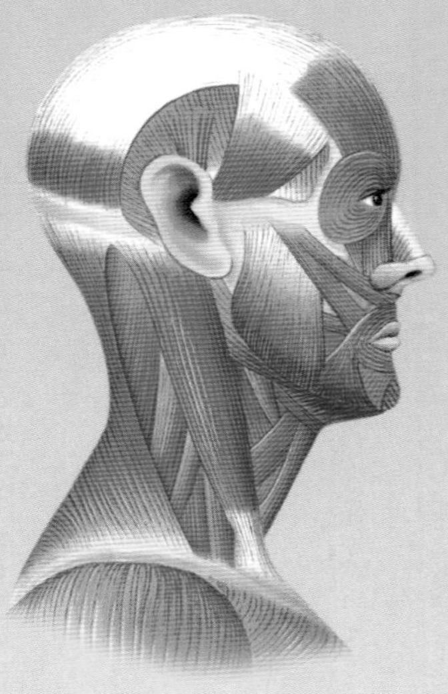

Muscles of the head and neck

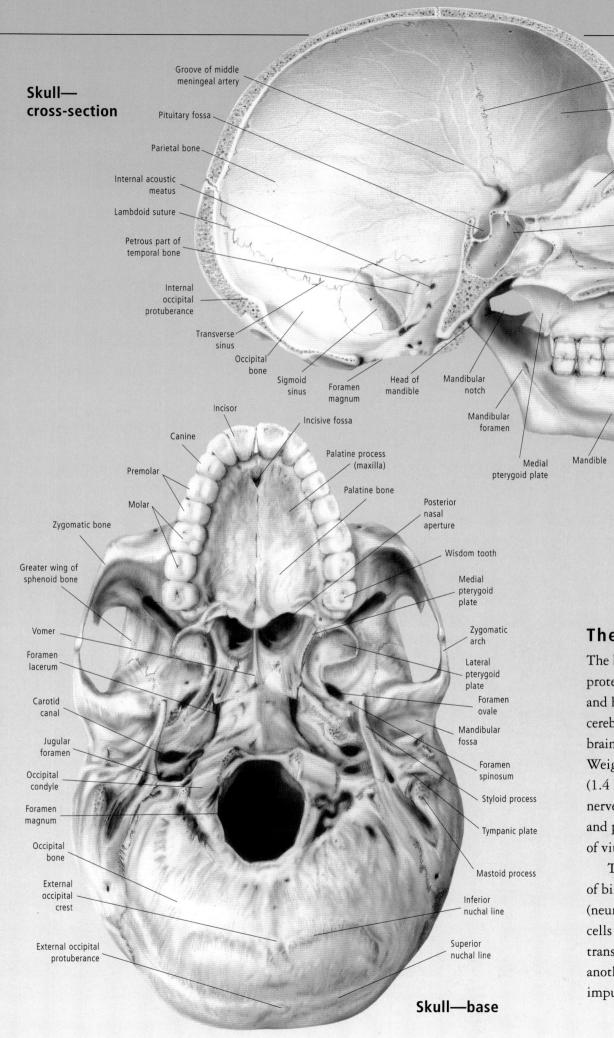

Skull—cross-section

Groove of middle meningeal artery

Pituitary fossa

Parietal bone

Internal acoustic meatus

Lambdoid suture

Petrous part of temporal bone

Internal occipital protuberance

Transverse sinus

Occipital bone

Sigmoid sinus

Foramen magnum

Head of mandible

Mandibular notch

Mandibular foramen

Medial pterygoid plate

Mandible

Bony palate

Coronal suture

Frontal bone

Crista galli (ethmoid)

Frontal sinus

Nasal bone

Sphenoid sinus

Perpendicular plate of ethmoid bone

Maxilla

Incisor

Incisive fossa

Canine

Palatine process (maxilla)

Premolar

Palatine bone

Molar

Posterior nasal aperture

Zygomatic bone

Wisdom tooth

Greater wing of sphenoid bone

Medial pterygoid plate

Vomer

Zygomatic arch

Foramen lacerum

Lateral pterygoid plate

Carotid canal

Foramen ovale

Jugular foramen

Mandibular fossa

Occipital condyle

Foramen spinosum

Foramen magnum

Styloid process

Occipital bone

Tympanic plate

External occipital crest

Mastoid process

External occipital protuberance

Inferior nuchal line

Superior nuchal line

Skull—base

The Brain

The brain lies within the protective casing of the skull, and has four main parts: the cerebrum, diencephalon, brain stem and cerebellum. Weighing about 3 pounds (1.4 kilograms), it is the nerve center of the body, and provides overall control of vital body functions.

The brain is made up of billions of nerve cells (neurons) and supporting cells (glia); these neurons transmit messages to one another by electrical impulses or chemical release.

THE CEREBRUM

The two hemispheres of the brain, joined together
by the corpus callosum, and covered by the gray
matter of the cerebral cortex, form the cerebrum.
Ridges and grooves on the surface of the cortex,
called gyri and sulci respectively, create its folded
appearance, which belies its large surface area;
it accounts for 40 percent of the brain mass,
and is the location of the highest level of
neural processing. Fissures and sulci divide
the cortex into separate functional areas,
known as lobes. The four lobes are named
according to the bone which overlies them,
that is, the frontal lobe, the parietal lobe,
the occipital lobe and the temporal lobe.
Each lobe contains many different areas
concerned with particular functions.

Beneath the gray matter of the cerebral cortex
is a thick mass of white matter, responsible for
transmitting information between different parts
of the cortex or between the cortex and other parts
of the brain. Within the white matter are isolated areas
of gray matter, known as basal ganglia, which play a part
in the control of movement.

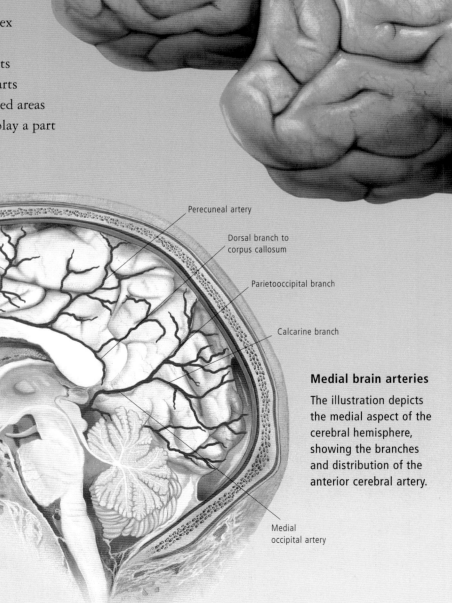

Paracentral artery

Pericallosal artery

Perecuneal artery

Medial
frontal
branches
— Posterior
Intermediate
Anterior

Dorsal branch to
corpus callosum

Callosomarginal
artery

Parietooccipital branch

Polar
frontal
artery

Calcarine branch

Right anterior
cerebral artery

Medial
frontobasal
artery

Medial
striate
artery

Medial brain arteries

The illustration depicts
the medial aspect of the
cerebral hemisphere,
showing the branches
and distribution of the
anterior cerebral artery.

Medial
occipital artery

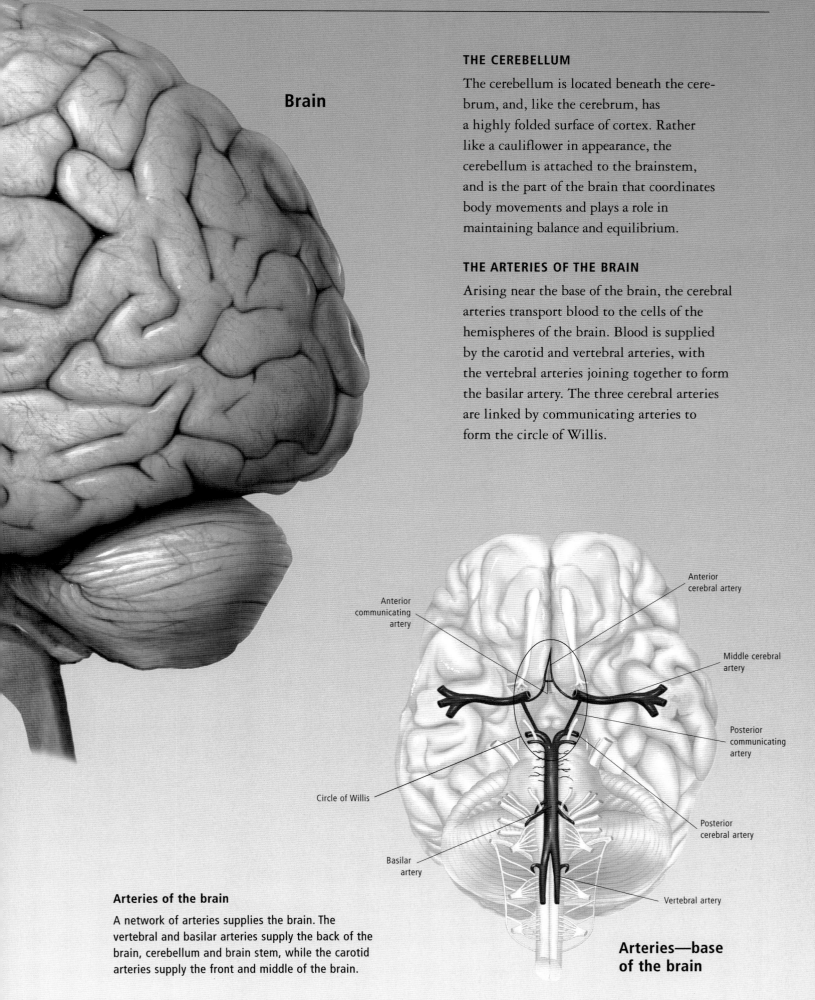

Brain

THE CEREBELLUM

The cerebellum is located beneath the cerebrum, and, like the cerebrum, has a highly folded surface of cortex. Rather like a cauliflower in appearance, the cerebellum is attached to the brainstem, and is the part of the brain that coordinates body movements and plays a role in maintaining balance and equilibrium.

THE ARTERIES OF THE BRAIN

Arising near the base of the brain, the cerebral arteries transport blood to the cells of the hemispheres of the brain. Blood is supplied by the carotid and vertebral arteries, with the vertebral arteries joining together to form the basilar artery. The three cerebral arteries are linked by communicating arteries to form the circle of Willis.

Anterior communicating artery

Anterior cerebral artery

Middle cerebral artery

Posterior communicating artery

Circle of Willis

Posterior cerebral artery

Basilar artery

Vertebral artery

Arteries of the brain

A network of arteries supplies the brain. The vertebral and basilar arteries supply the back of the brain, cerebellum and brain stem, while the carotid arteries supply the front and middle of the brain.

Arteries—base of the brain

The Diencephalon

Located beneath the cerebral hemispheres, the diencephalon has two main structures: the thalamus and the hypothalamus. The thalamus relays sensory information to the cerebral cortex and controls motor activity, while the hypothalamus serves as an interface between the brain and the autonomic nervous system. It controls a wide range of body functions, including eating, drinking, sexual function and body temperature, and also plays a role in expressing emotions.

The Brain Stem

The brain stem consists of the midbrain (also known as the mesencephalon), the pons and the medulla. As a continuation of the spinal cord, the brain stem relays sensory information from the spinal cord to the brain along ascending pathways, and sends motor information from the cortex to the spinal cord along descending pathways. The brain stem contains many important reflex centers which control vital functions such as heartbeat and respiration. The midbrain is important in controlling eye functions, while the medulla is involved in regulating sleep and arousal, and in pain perception.

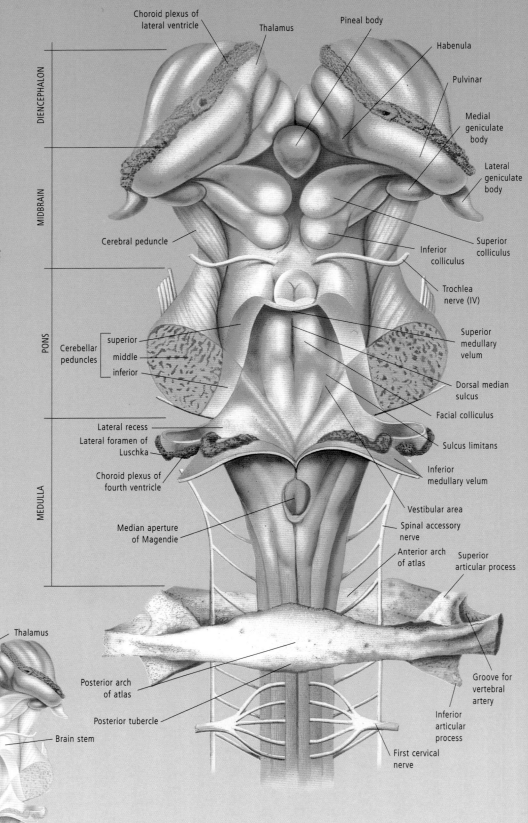

Thalamus

The thalamus is part of the diencephalon, and is a relay center for motor and sensory information. Messages are transmitted, via the thalamus, from the spinal cord to the cerebral cortex and cerebellum.

Brain Stem

The brain stem is the link between the spinal cord and the cerebral cortex. Continuous with the spinal cord below it, the brain stem has three parts, the midbrain, pons and medulla. The brain stem carries messages between the spinal cord and the brain, and centers within the brain stem regulate many vital functions including breathing, heartbeat and blood pressure.

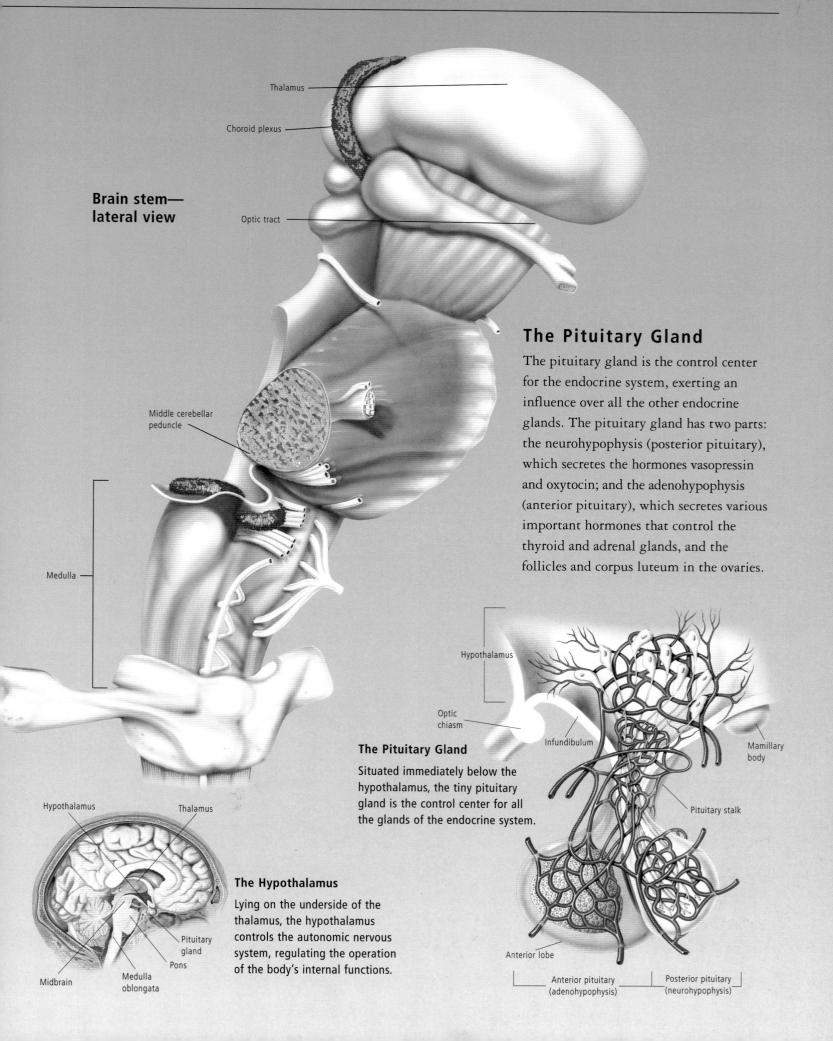

Thalamus

Choroid plexus

Brain stem— lateral view

Optic tract

Middle cerebellar peduncle

Medulla

The Pituitary Gland

The pituitary gland is the control center for the endocrine system, exerting an influence over all the other endocrine glands. The pituitary gland has two parts: the neurohypophysis (posterior pituitary), which secretes the hormones vasopressin and oxytocin; and the adenohypophysis (anterior pituitary), which secretes various important hormones that control the thyroid and adrenal glands, and the follicles and corpus luteum in the ovaries.

Hypothalamus

Optic chiasm

Infundibulum

Mamillary body

Pituitary stalk

The Pituitary Gland

Situated immediately below the hypothalamus, the tiny pituitary gland is the control center for all the glands of the endocrine system.

Hypothalamus

Thalamus

The Hypothalamus

Lying on the underside of the thalamus, the hypothalamus controls the autonomic nervous system, regulating the operation of the body's internal functions.

Pituitary gland

Pons

Midbrain

Medulla oblongata

Anterior lobe

Anterior pituitary (adenohypophysis)

Posterior pituitary (neurohypophysis)

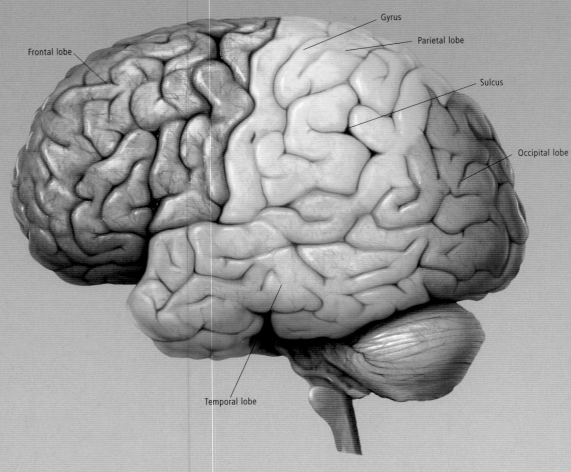

Frontal lobe

Gyrus

Parietal lobe

Sulcus

Occipital lobe

Temporal lobe

Lobes of the brain

The heavily folded surface of the cerebral cortex (cerebrum) creates ridges and grooves. A ridge is called a gyrus; a groove is called a sulcus if shallow, or a fissure if deep. The cortex is divided into four separate functional areas called lobes.

Functional areas

Certain areas of the cerebral cortex are associated with particular functions. For example, the postcentral gyrus (sensory cortex) is associated with sensations from skin, muscles and joints. The precentral gyrus (motor cortex) is associated with the voluntary control of skeletal muscles.

THE LOBES OF THE BRAIN

The cerebral cortex is divided into four lobes by fissures and sulci. The four lobes are the frontal lobe, the parietal lobe, the occipital lobe and the temporal lobe. Within each lobe are various regions involved in processing motor and sensory information.

THE FUNCTIONAL AREAS OF THE BRAIN

Various regions of the cerebral cortex are responsible for certain functions. Areas of the temporal lobe are associated with hearing and memory while areas of the occipital lobe are associated with vision.

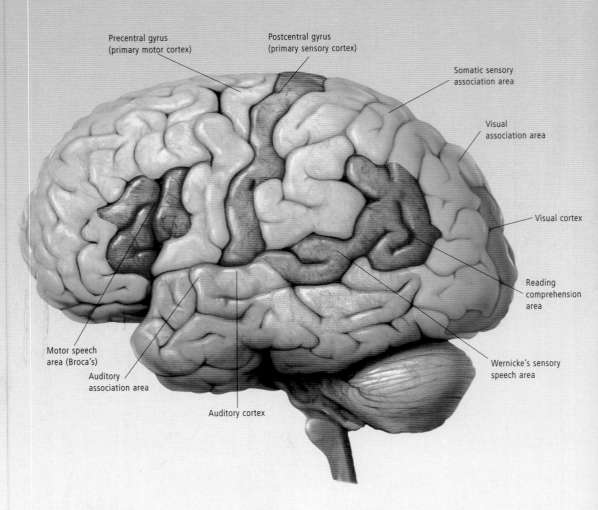

Precentral gyrus (primary motor cortex)

Postcentral gyrus (primary sensory cortex)

Somatic sensory association area

Visual association area

Visual cortex

Reading comprehension area

Motor speech area (Broca's)

Auditory association area

Auditory cortex

Wernicke's sensory speech area

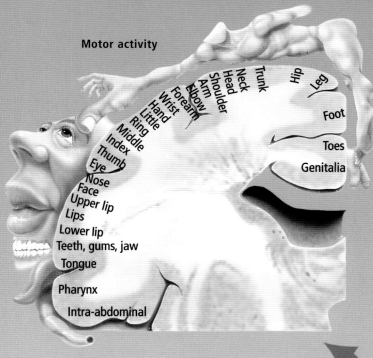

Motor activity

Leg
Hip
Trunk
Neck
Head
Shoulder
Arm
Elbow
Forearm
Wrist
Hand
Little
Ring
Middle
Index
Thumb
Eye
Nose
Face
Upper lip
Lips
Lower lip
Teeth, gums, jaw
Tongue
Pharynx
Intra-abdominal
Foot
Toes
Genitalia

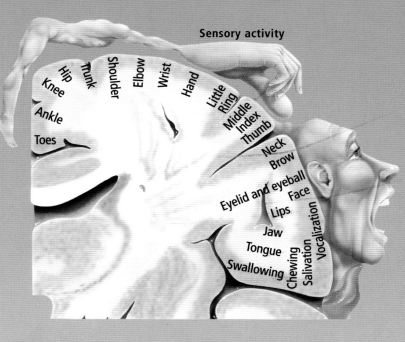

Sensory activity

Hip
Knee
Trunk
Shoulder
Ankle
Elbow
Wrist
Hand
Little
Ring
Middle
Index
Thumb
Toes
Neck
Brow
Eyelid and eyeball
Face
Lips
Jaw
Tongue
Swallowing
Chewing
Salivation
Vocalization

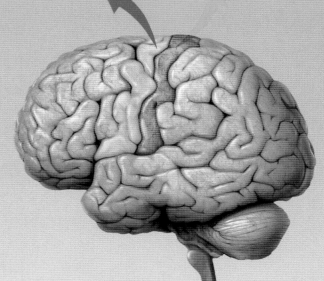

Organization of the motor and sensory areas of the cerebral cortex

The precentral gyrus (top left) is involved in motor activity in particular parts of the body, with the relative size of the body parts indicating the amount of involvement. Similarly, the postcentral gyrus (top right) is involved in sensory activity in particular parts of the body, and the degree of involvement is again indicated by the proportionate sizing.

THE MOTOR AND SENSORY AREAS OF THE CEREBRAL CORTEX

The precentral gyrus (motor cortex) is associated with the voluntary control of skeletal muscles. The postcentral gyrus (sensory cortex) is associated with sensations from skin, muscles and joints.

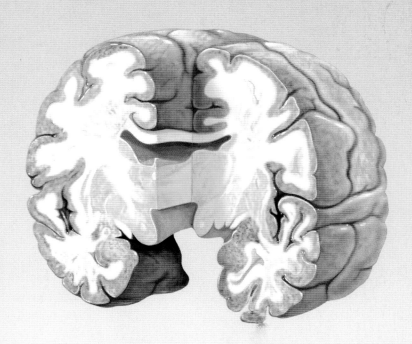

Cerebral cortex

The cerebral cortex is the highly folded outer layer of the brain; it is made up of nerve cells (neurons). Most messages from the brain originate in the cerebral cortex, and this is where the more complex functions of the brain such as thinking, decision-making, speaking and hearing take place.

THE BRAIN VENTRICLES

The four ventricles, or cavities, of the brain are the site of cerebrospinal fluid (CSF) production. As the ventricles connect to the central canal of the spinal cord and the subarachnoid space around the brain, the CSF fills the ventricles and flows through these connecting structures, draining into the venous system. Produced continuously, the total volume of CSF is replaced several times daily.

THE LIMBIC SYSTEM

The key elements of the limbic system are the hippocampus, amygdala, septal area and hypothalamus. These interconnected structures are involved in behaviors associated with survival, expression of emotion and the formation of memory. Each of the components plays a role in the overall limbic system. The hippocampus and amygdala are both located in the temporal lobe; the hippocampus is involved in the formation of new memories, while the amygdala is concerned with expression of emotion. The septal area lies in the inner surface of the brain and is thought to be associated with pleasure or reward. The hypothalamus controls hormone production, and its influence over the body results in the physical reactions associated with emotions, such as raised blood pressure, or quickening of the heart and breathing rate when anxious.

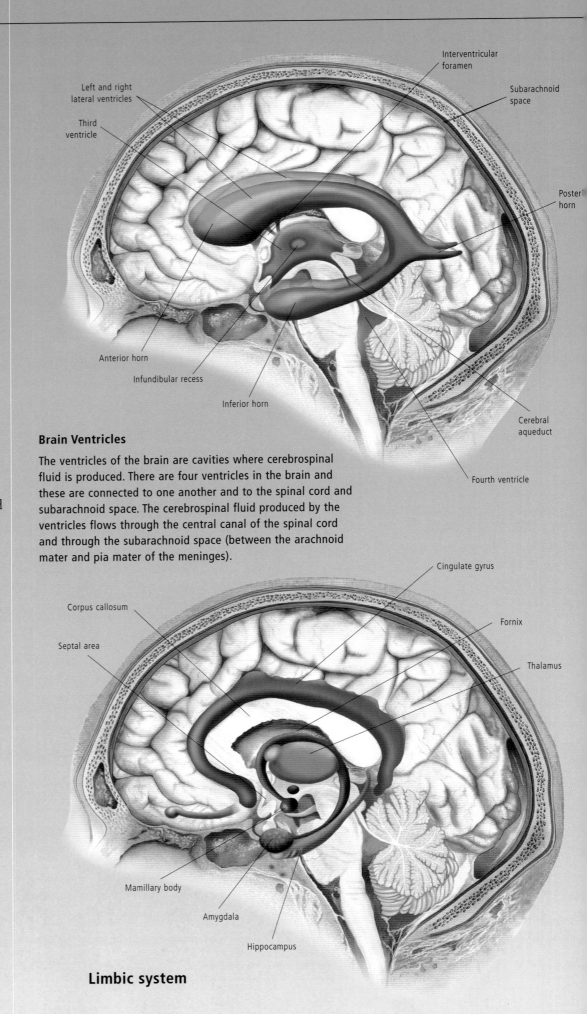

Brain Ventricles

The ventricles of the brain are cavities where cerebrospinal fluid is produced. There are four ventricles in the brain and these are connected to one another and to the spinal cord and subarachnoid space. The cerebrospinal fluid produced by the ventricles flows through the central canal of the spinal cord and through the subarachnoid space (between the arachnoid mater and pia mater of the meninges).

Limbic system

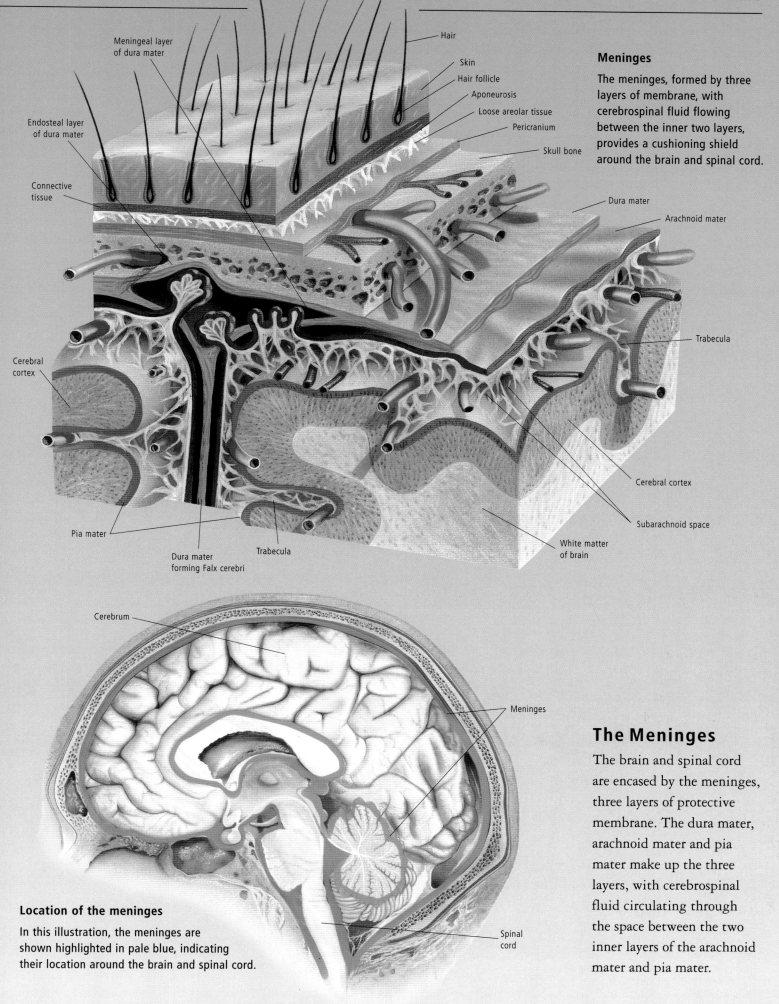

Meningeal layer
of dura mater

Hair

Skin

Hair follicle

Aponeurosis

Loose areolar tissue

Pericranium

Skull bone

Endosteal layer
of dura mater

Connective
tissue

Dura mater

Arachnoid mater

Trabecula

Cerebral
cortex

Cerebral cortex

Subarachnoid space

Pia mater

White matter
of brain

Dura mater
forming Falx cerebri

Trabecula

Meninges

The meninges, formed by three
layers of membrane, with
cerebrospinal fluid flowing
between the inner two layers,
provides a cushioning shield
around the brain and spinal cord.

Cerebrum

Meninges

Spinal
cord

The Meninges

The brain and spinal cord
are encased by the meninges,
three layers of protective
membrane. The dura mater,
arachnoid mater and pia
mater make up the three
layers, with cerebrospinal
fluid circulating through
the space between the two
inner layers of the arachnoid
mater and pia mater.

Location of the meninges

In this illustration, the meninges are
shown highlighted in pale blue, indicating
their location around the brain and spinal cord.

General Senses

Our bodies experience various sensations, such as touch, pain, vibration, temperature and pressure. Nerve endings, known as receptors, send impulses along sensory nerves to the spinal cord. These impulses then travel up the spinal cord along special pathways through the thalamus to the cerebral cortex.

Many of the general sense receptors are found in the skin, with concentrations of mechanoreceptors in certain areas such as the fingertips, lips, palms, toes, nipples, the glans of the penis and the clitoris.

Thermoreceptors detect temperature, with separate receptors for heat and cold. Found throughout the body, they are concentrated in areas such as the lips, mouth and anus.

Proprioreceptors provide information on joint, tendon and muscle status; this information allows coordination of muscle movements.

Pain receptors, known as nocireceptors, are widely distributed throughout the body tissues, being especially common in the skin and joints. These pain receptors are also able to detect pain in the internal organs of the body.

Special Senses

Our special senses are sight, smell (olfaction), taste, hearing and equilibrium (balance). The organs for the special senses, the eyes, nose, mouth and ears, each have nerve endings called receptors, tailored to their specific function. Sight is detected by photoreceptors; touch, sound and equilibrium are detected by mechanoreceptors; and smell and taste are detected by chemoreceptors. These receptors relay their information via cranial nerves to the brain for interpretation.

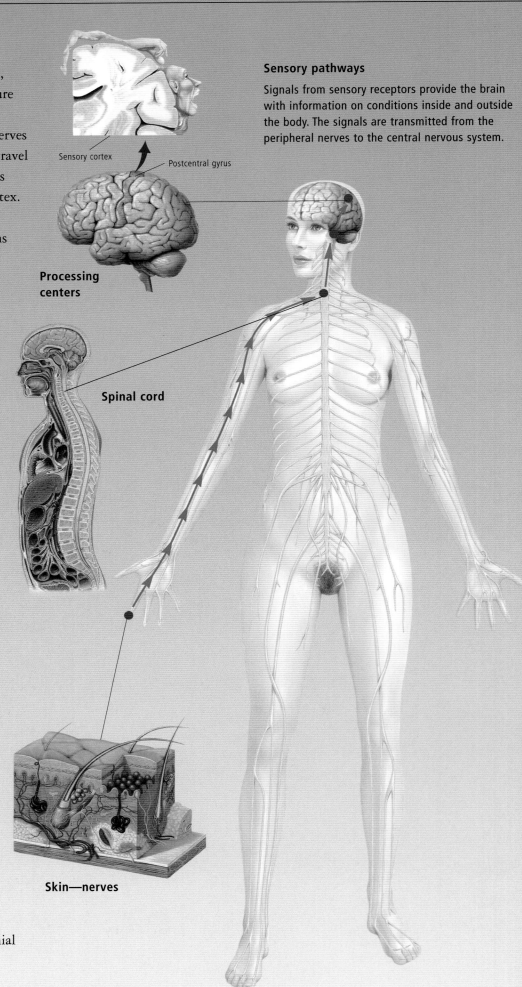

Sensory cortex

Postcentral gyrus

Processing centers

Spinal cord

Skin—nerves

Sensory pathways

Signals from sensory receptors provide the brain with information on conditions inside and outside the body. The signals are transmitted from the peripheral nerves to the central nervous system.

Sight

Light-sensitive photoreceptors in the eyes, called rods and cones, transmit information along the optic nerve. This information is relayed to the occipital cortex of the brain for interpretation.

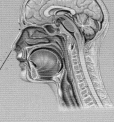

Smell

The nasal cavity contains chemo-receptors, capable of detecting thousands of odors. The chemo-receptors transmit signals via the olfactory nerve to olfactory areas of the brain, where they are interpreted as smell.

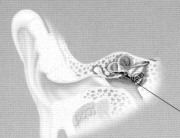

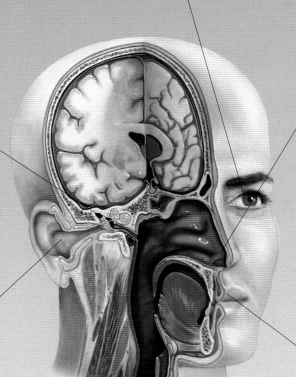

Balance

Information on balance is transmitted through the vestibulocochlear nerve by the vestibular system, comprising the semicircular canals, utricle and saccule of the ear.

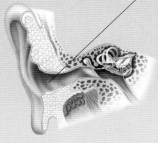

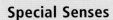

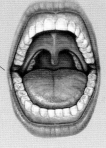

Taste

Taste buds (chemoreceptors) are located on the tongue, palate and throat. Information on salty, sweet, sour and bitter flavors is sent via the cranial nerves to the brain, where the taste is recognized. The sense of taste is enhanced by the sense of smell.

Hearing

Mechanoreceptors in the cochlea in the ear convert sound waves into impulses. These impulses are sent to the brain via the vestibulocochlear nerve, providing the brain with information on sound.

Special Senses

The special senses are sight, smell, taste, hearing and equilibrium (balance). Nerve impulses sent from the special sense organs travel to the brain via the cranial nerves.

The Face

The facial bones are the foundation upon which our appearance is built, with the overlying skin and muscles adding the finishing touches, allowing us to recognize one another. Our facial expressions, conveying much of what we think and feel, are produced by the facial muscles and nerves. The special sense organs (eyes, ears, nose and tongue) are all part of the facial structure.

The face is made up of 14 facial bones, muscles, skin, eyes, nose, jaws, cheeks and chin, as well as the nerves and blood vessels supplying these structures. The muscles of the face include a circular muscle around both the eyes and the mouth. The mouth and cheek muscles are important in speech; the jaw muscles operate to move the lower jaw when eating and speaking.

The Cranial Nerves

Arising mainly from the brain stem, the 12 pairs of cranial nerves innervate the muscles and sensory structures of the head and neck (including skin, membranes, eyes and ears). They also distribute nerves to the organs of the chest (trachea, bronchi, lungs and heart).

Cranial nerves

Olfactory nerve (I)

The first cranial nerve is concerned with the sense of smell. Nerve fibers starting in the mucous membranes of the nose carry messages to the cerebrum.

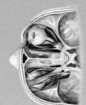

Optic nerve (II)

Visual impulses from the retina are sent along the optic nerve (the second cranial nerve) to the brain.

Oculomotor (III), trochlear (IV) and abducent (VI) nerves

These cranial nerves control movement of the muscles which move the eyeball and eyelids, and allow focusing.

Spinal accessory nerve (XI)

The eleventh cranial nerve is primarily responsible for movement of the muscles of the upper shoulders, head, neck, and larynx and pharynx.

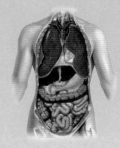

Vagus nerve (X)

The tenth cranial nerve is involved with functions such as coughing, sneezing, swallowing, speaking, secretions from the glands of the stomach, as well as the sensation of hunger.

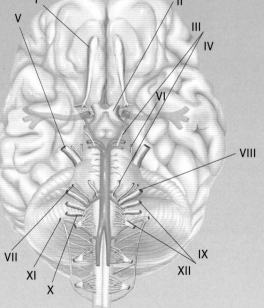
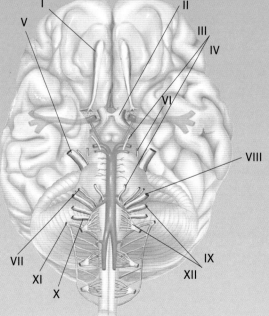

Trigeminal nerve (V)

The trigeminal nerve (fifth cranial nerve) has three sections: the ophthalmic, maxillary and mandibular divisions. They supply sensory fibers to the forehead and skin of the cheek, and join motor fibers to the muscles used for chewing.

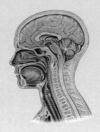

Glossopharyngeal (IX) and hypoglossal (XII) nerves

Supplying the carotid sinus, the ninth cranial nerve is responsible for the reflex control of the heart. It also supplies the back part of the tongue and the soft palate. The twelfth cranial nerve controls movement of the tongue.

Vestibulocochlear nerve (VIII)

Located behind the facial nerve, the eighth cranial nerve carries impulses for the sense of balance.

Facial nerve (VII)

The facial nerve is the seventh cranial nerve. It provides the motor fibers for facial expression. It is also responsible for the sensation of taste in the front part of the tongue.

Face

A complex range of thoughts and emotions is expressed in the face—this requires an intricate system of muscles and nerves. The special sense organs (eyes, nose, ears and tongue) are all part of the facial structure.

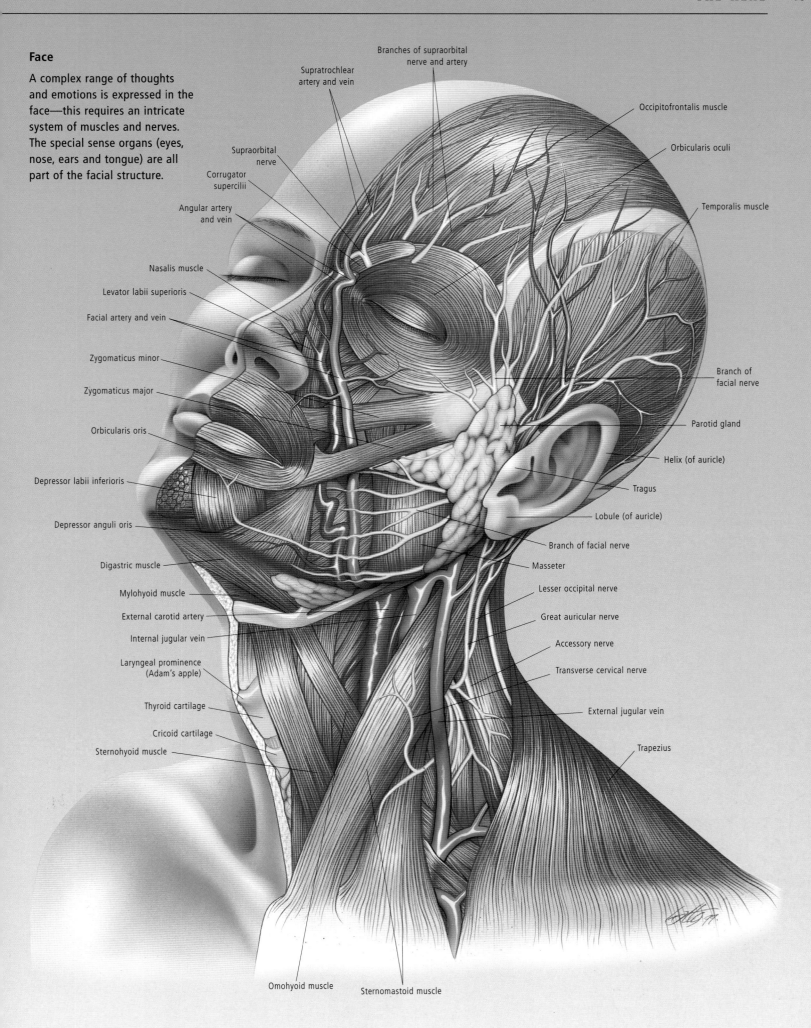

Supratrochlear artery and vein

Branches of supraorbital nerve and artery

Occipitofrontalis muscle

Orbicularis oculi

Temporalis muscle

Supraorbital nerve

Corrugator supercilii

Angular artery and vein

Nasalis muscle

Levator labii superioris

Facial artery and vein

Zygomaticus minor

Zygomaticus major

Orbicularis oris

Depressor labii inferioris

Depressor anguli oris

Digastric muscle

Mylohyoid muscle

External carotid artery

Internal jugular vein

Laryngeal prominence (Adam's apple)

Thyroid cartilage

Cricoid cartilage

Sternohyoid muscle

Branch of facial nerve

Parotid gland

Helix (of auricle)

Tragus

Lobule (of auricle)

Branch of facial nerve

Masseter

Lesser occipital nerve

Great auricular nerve

Accessory nerve

Transverse cervical nerve

External jugular vein

Trapezius

Omohyoid muscle

Sternomastoid muscle

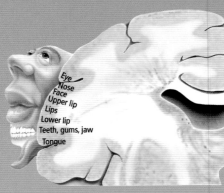

Motor

Eye
Nose
Face
Upper lip
Lips
Lower lip
Teeth, gums, jaw
Tongue

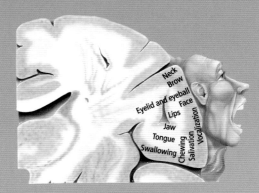

Sensory

Neck
Brow
Eyelid and eyeball
Face
Lips
Jaw
Tongue
Swallowing
Chewing
Salivation
Vocalization

The Facial Nerves

The facial nerves on either side of the face control the facial muscles, supply the salivary glands of the mouth and the lacrimal glands of the eye, and carry taste sensations from the front two-thirds of the tongue. Arising at the brain stem, the facial nerves lie in front of each of the ears, branching out across each side of the face.

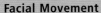
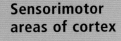

Precentral gyrus

Postcentral gyrus

Sensorimotor areas of cortex

Facial Movement

The motor activities of the face, such as smiling, are processed by the precentral gyrus, while the sensory activities associated with the facial skin are processed by the postcentral gyrus. In the illustration, the size of the various parts of the face indicates the proportion of influence exerted over each part by the gyri.

Superior temporal line

Temporalis muscle

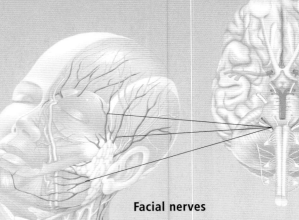

Facial nerves

The facial nerves fan out across the face in front of the ear. They control the muscles of the face, supply the salivary glands, and transmit taste sensations from the front portion of the tongue.

Surface muscles of the jaw

The masseter muscle is used in biting and chewing actions (mastication).

The Jaw

The temporomandibular joint joins the mandible and maxillae to form the jaw, with the mandible forming the lower moveable part, and the maxillae forming the upper fixed part. The maxillae form the upper jaw; this has an alveolar part, containing the sockets for the upper teeth, a palatine part (the hard palate in the roof of the mouth), and a hollow body. The mandible has a thickened body, an alveolar part (containing sockets for the lower teeth) and a ramus. At the top of the ramus is a small rounded head which fits into a socket on the base of the skull. This forms the temporomandibular joint, connecting

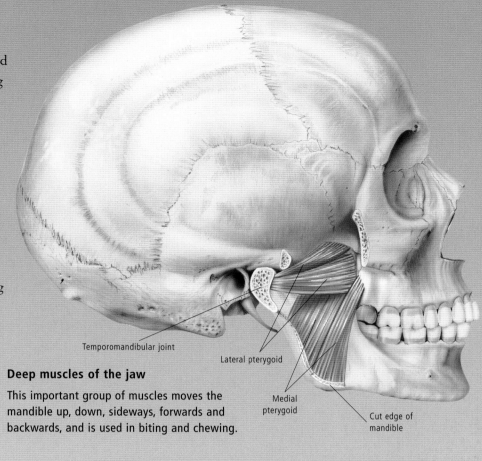

Temporomandibular joint

Lateral pterygoid

Medial pterygoid

Cut edge of mandible

Deep muscles of the jaw

This important group of muscles moves the mandible up, down, sideways, forwards and backwards, and is used in biting and chewing.

the mandible and the maxillae. The joint has a reinforcing capsule and is supported by strong ligaments, allowing it to perform both gliding and hinge movements.

The muscles of mastication, including the medial and lateral pterygoid, masseter and temporalis muscles, move the joint, working to close the jaw by moving the mandible upwards. Both the mandible and maxillae contain sockets for the teeth. In an adult, each of the two parts of the jaw contains sockets for 16 teeth.

Zygomatic arch

Temporomandibular joint

Masseter muscle

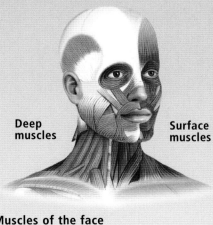

Deep muscles

Surface muscles

Muscles of the face

The muscles of the face, in conjunction with the nerves, create the facial expressions that convey our thoughts and feelings. The muscles of the mouth and cheeks play a key role in both speaking and eating; the muscles around the eye open and close the eye and perform blinking movements.

The Eye

Our "window to the world", the eye is a remarkable organ. The organ of sight, this complex structure relays images to the brain for interpretation. The three layers that make up the eye are: the outer layer of the sclera and cornea, the middle layer of the uvea and lens, and the inner layer of the retina.

Forming the spherical shape of the eye, the sclera or "white of the eye", is made up of tough fibrous tissue. The sclera is the outer layer of the eye, providing the attachment point for the muscles that move the eye. At the front part of this outer layer is the transparent cornea, and at the back is the exit point for the optic nerve. The cornea bends the light reflected off a seen object, and in conjunction with the lens, forms the image on the retina.

The middle layer of the eye, the uvea, has three components: the choroid, ciliary body and iris. At the back of the uvea is the choroid, coursing with blood vessels and nerves supplying the cornea, ciliary body and iris. Towards the front of the uvea, the choroid becomes the ciliary body, which joins with the iris. The ciliary muscles hold the lens of the eye in a supportive mesh of tiny, thread-like fibers, called the zonule. The center of the iris is the pupil, with the muscles around the iris adjusting the pupil size, dilating or reducing the size, depending on the light level. This function, along with the movements of the ciliary muscles is under the control of the autonomic nervous system. The space

> **DID YOU KNOW?**
>
> The skin of the eyelids is soft and flexible, enabling it to perform the blinking movements necessary to lubricate the eye. On average, a blink lasts around 50–100 milliseconds.

Eye

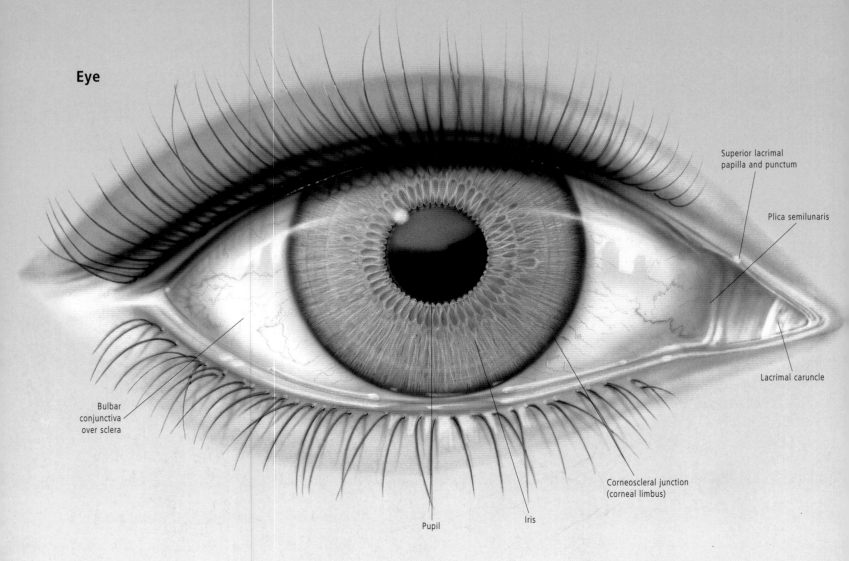

Superior lacrimal papilla and punctum

Plica semilunaris

Lacrimal caruncle

Bulbar conjunctiva over sclera

Corneoscleral junction (corneal limbus)

Pupil

Iris

Eyeball

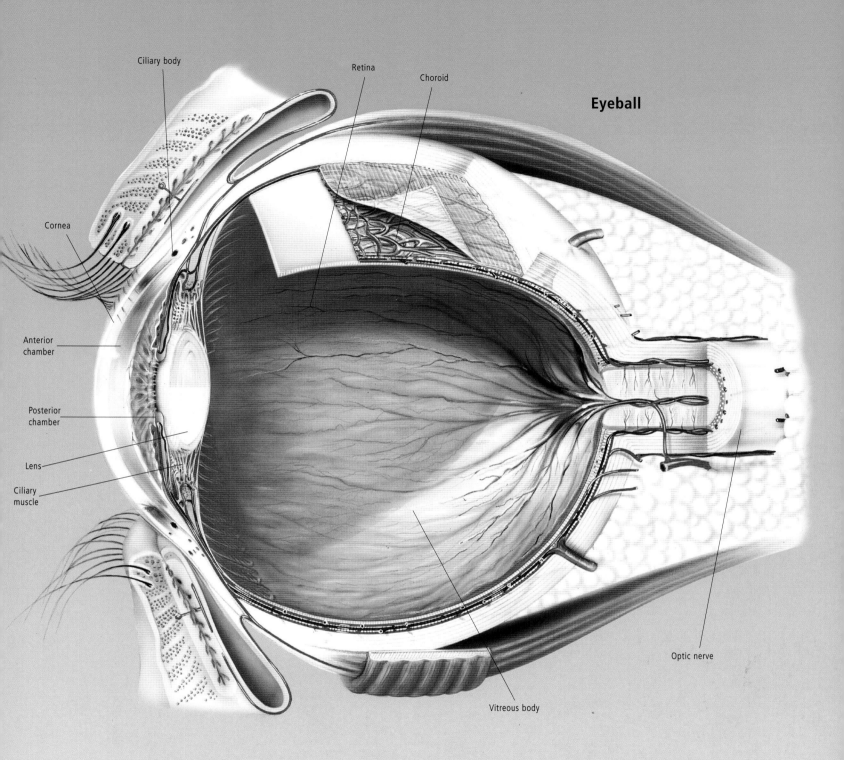

Ciliary body

Retina

Choroid

Cornea

Anterior chamber

Posterior chamber

Lens

Ciliary muscle

Optic nerve

Vitreous body

in front of the lens and zonule is divided into the anterior and posterior chambers. To nourish the lens and cornea, the ciliary body secretes a fluid called aqueous humor into the posterior chamber; this then passes through the pupil into the anterior chamber. The cavity behind the lens is filled with a gel, known as the vitreous body, which helps to maintain the shape of the eyeball.

The retina is the inner layer of the eye, made up of light sensitive cells (photoreceptors) called rods and cones, with more than 100 million rods and over 5 million cones. Light passes through the cornea, pupil, lens and vitreous humor to reach the retina. Images relayed to the retina are transported via the optic nerve, with the nerves of each eye converging at the optic chiasm, and connecting to the brain.

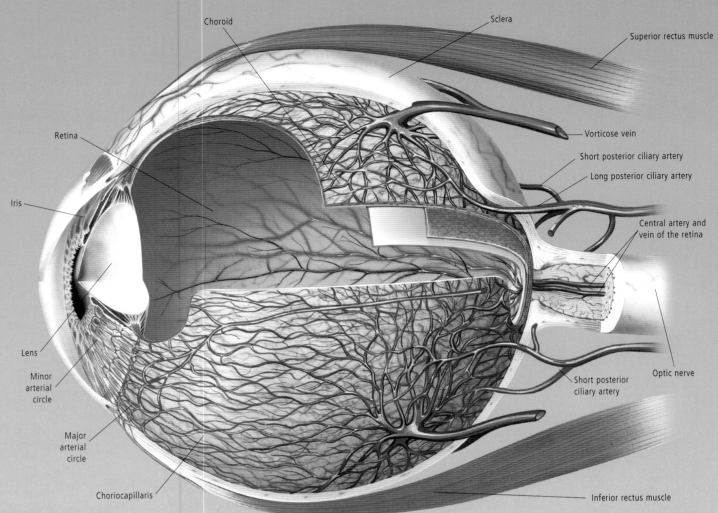

Choroid

Sclera

Superior rectus muscle

Retina

Vorticose vein

Short posterior ciliary artery

Long posterior ciliary artery

Iris

Central artery and vein of the retina

Lens

Minor arterial circle

Short posterior ciliary artery

Optic nerve

Major arterial circle

Choriocapillaris

Inferior rectus muscle

Blood vessels of the eye

A complex system of arteries, veins and capillaries
supply the eye. The central artery and vein of
the eye run through the optic nerve, with their
branches spreading out across the retinal surface.
The ciliary arteries also supply the eye.

THE BLOOD VESSELS OF THE EYE

The central artery and vein of the retina run through the optic
nerve, subdividing into four main branches, which subdivide
further, becoming a network of capillaries.

THE EYELASHES AND EYELIDS

Both the eyelashes and eyelids serve to protect the eye from
glare and dust. The eyelids are folds of skin that close to cover
the surface of the eye. The undersurface of the eyelid is covered
by a continuation of the conjunctiva, a mucous membrane
covering the surface of the eyeball, while the edges of the eyelids
contain oil-secreting glands (meibomian glands) that lubricate
the eyelids. Adding further protection to the delicate sight organ
are the eyelashes, which extend out from the eyelids.

MOVEMENTS OF THE EYEBALL

The movement of the eyeball is undertaken by three pairs of
muscles. Finely tuned movements allow the eye to look up,
down, left and right, providing a wide field of vision.

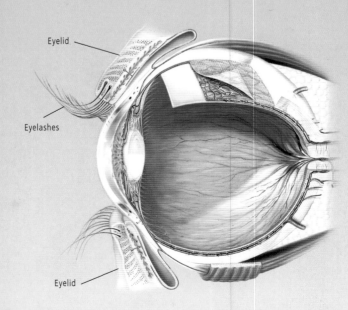

Eyelid

Eyelashes

Eyelid

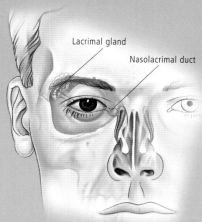

Lacrimal gland

Nasolacrimal duct

THE LACRIMAL APPARATUS

The lacrimal glands are situated at the upper outer corner of the orbit, the cavity at the front of the skull housing the eyeball. Tears are secreted by the lacrimal gland; these tears form a protective film and lubricate the eye. The blinking movement of the eyelid moves the fluid toward its drainage point, the nasolacrimal duct, which opens to the nose.

Lacrimal glands

The lacrimal gland produces tears to keep the eye moist, as well as providing lubrication and protection against infection. As the eyelid blinks, tears move across the surface of the eye, toward their exit point at the lower inner corner of the eye, the nasolacrimal duct.

DID YOU KNOW?

When we cry, the amount of tears produced by the lacrimal glands is too great for the nasolacrimal duct to manage, so the tears spill over.

Movements of the eyeball

The eyeball lies in a custom-made cavity in the skull, the orbit. The six muscles supporting the eyeball in the orbit work together in three pairs, controlling the directional movement of the eyeball. Carefully co-ordinated movements and calibrations, requiring the participation of all six muscles, move the eye up, down, left and right. This great range of movement provides our wide field of vision.

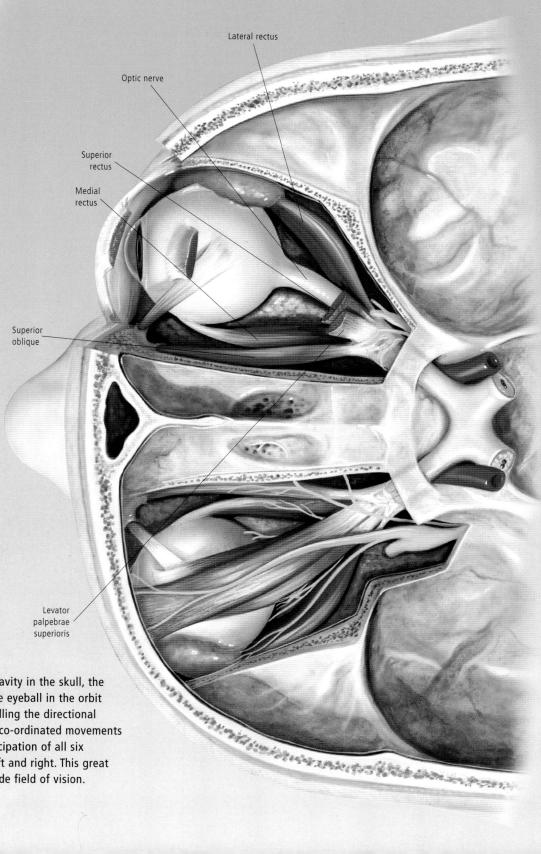

Lateral rectus

Optic nerve

Superior rectus

Medial rectus

Superior oblique

Levator palpebrae superioris

How we see

The structure of the eye is specially designed to bend and concentrate light rays to form a tiny image of a seen object on the back of the eye. This is then transported to the brain in the form of nerve impulses. Light rays entering the eye strike the cornea, which bends (refracts) the rays bringing them closer together. The rays then pass through the lens which focuses the rays on the back of the retina. The retina consists of a layer of light-sensitive cells—rods and cones. When stimulated by light, the rods and cones send electrical signals along the cells of the optic nerves to the brain.

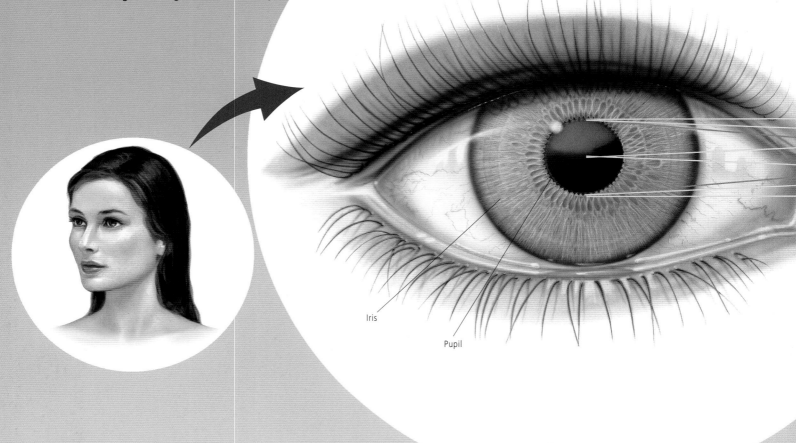

Iris

Pupil

Sight

Our left and right eyes have slightly different fields of vision, which overlap and merge together to focus on a seen object, allowing us to discern distance and 3-D structure. This is known as binocular vision.

When we look at an object, the light rays reflected off the object enter the eye. The amount of light entering the eye is governed by the iris, the colored part of the eye which lies behind the cornea. Light rays passing through the cornea are adjusted to bring the rays closer together; these rays then pass through the pupil to the lens. Muscular action can alter the shape of the lens to allow focus on distant and near objects. The light rays then pass through the vitreous humor to the retina. As the light rays travel through the eye to the retina they are concentrated and bent (refracted), so the image received on the back of the retina is an inverted, transposed version of the seen object. The photoreceptors

of the retina, called rods and cones, are activated by the light rays and send nerve impulses along the cells of the optic nerve.

The optic nerve (cranial nerve II) transmits the impulses to the thalamus, where some processing of the visual information is conducted by the lateral geniculate nuclei of the thalamus. The visual information is then relayed to the visual cortex in the occipital lobe of the brain for interpretation.

The visual cortex interprets and makes sense of the nerve impulses sent from the eyes via the optic nerves and thalamus. This is where the messages from the right and left eyes are merged to form one image, and the seen object is recognized. The visual association cortex processes more complex features of the visual stimulus, such as color and movement. The images received on the retina are upside down, but the brain automatically corrects these images.

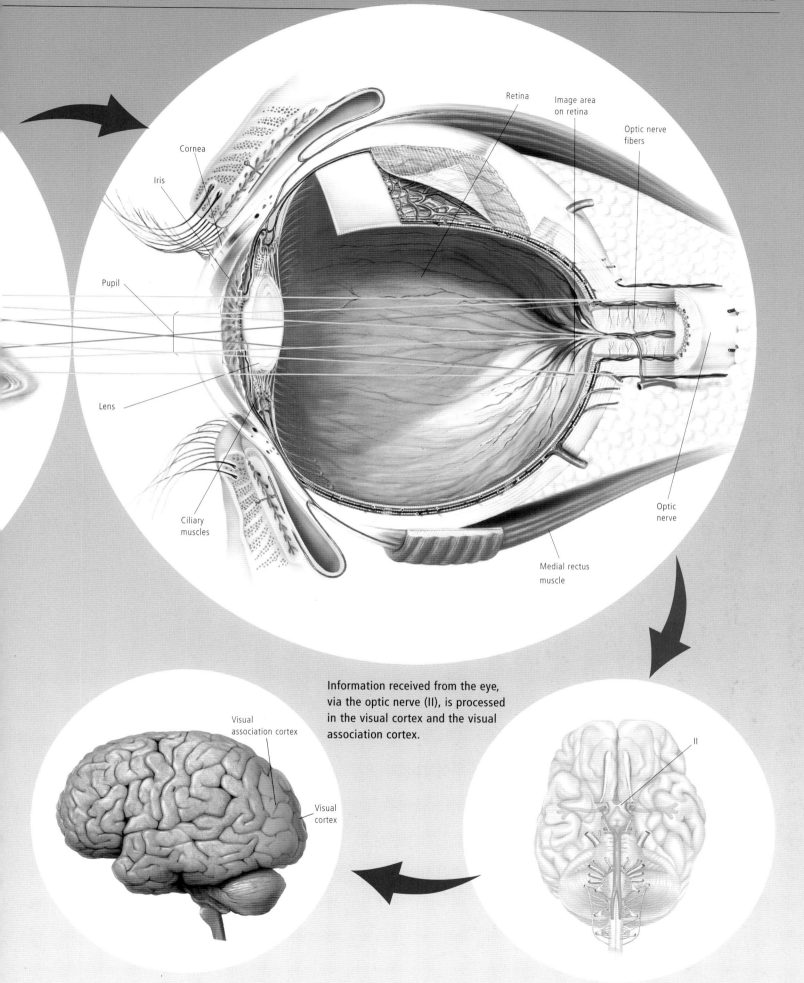

Cornea

Iris

Pupil

Lens

Ciliary
muscles

Retina

Image area
on retina

Optic nerve
fibers

Optic
nerve

Medial rectus
muscle

Information received from the eye,
via the optic nerve (II), is processed
in the visual cortex and the visual
association cortex.

Visual
association cortex

Visual
cortex

II

The Ear

The special sense organ of the ear is responsible for hearing and balance. Located in a hollow space in the temporal bone of the skull, the ear is made up of the outer ear, the middle ear and the inner ear.

Sound waves entering the ears are converted into mechanical vibrations and then into nerve impulses. The nerve impulses are then transmitted to the brain for interpretation. The ear also senses the body's position relative to gravity, sending information to the brain that allows the body to maintain postural equilibrium.

THE STRUCTURE AND FUNCTION OF THE EAR

The ear is positioned in a cavity of the temporal bone. The outer ear consists of the pinna or auricle and the auditory canal. The auditory canal is lined with glands that secrete wax (cerumen); the purpose of the wax is to trap dust and dirt. This canal connects the external ear to the eardrum.

The middle ear contains the ossicles, three tiny bones comprising the malleus (hammer), incus (anvil) and stapes (stirrup). The ossicles connect across the tympanic cavity to the oval window in the cochlea. Connecting the middle ear to the throat is the eustachian (auditory) tube.

The inner ear (labyrinth) contains the cochlea, the main organ of hearing, and the semicircular canals, the organs of balance. The vibrations of the ossicles travel to the cochlea, where they cause waves in the cochlear fluid. These vibrations trigger the receptors lodged in the organ of Corti, which sends nerve impulses along the vestibulo-cochlear nerve, the eighth cranial nerve, to the auditory cortex in the temporal lobe of brain.

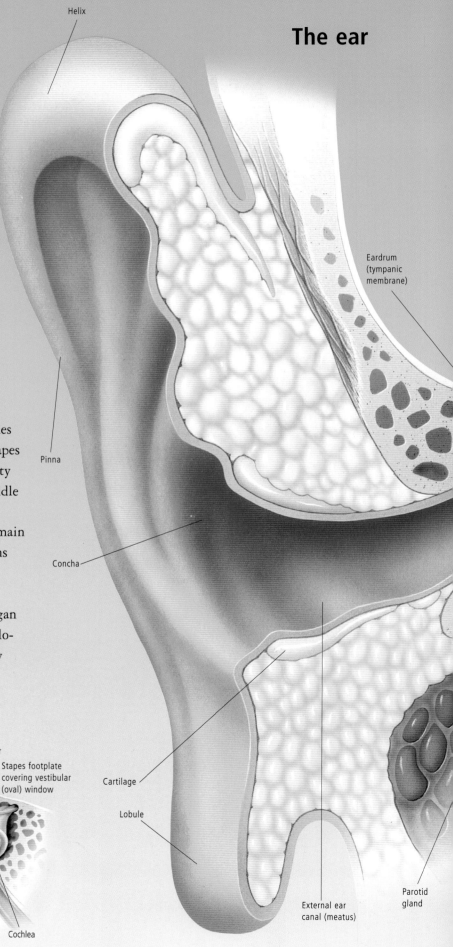

The ear

Helix

Eardrum (tympanic membrane)

Pinna

Concha

Cartilage

Lobule

External ear canal (meatus)

Parotid gland

Ossicles and semicircular canals

Ossicles

Malleus Incus Stapes

Semicircular canals

Stapes footplate covering vestibular (oval) window

Eardrum

Cochlea

The Eardrum

The eardrum (or tympanic membrane) is a thin membrane between the outer and middle ear, separating the two areas. Sound waves reverberate on the tympanic membrane, causing it to vibrate and create vibrations through the middle ear.

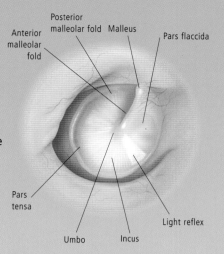

Anterior malleolar fold
Posterior malleolar fold
Malleus
Pars flaccida
Pars tensa
Umbo
Incus
Light reflex

Ageing of the ear

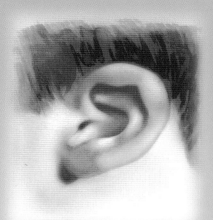

Child

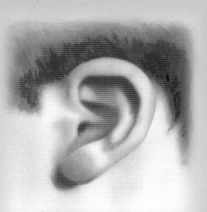

Adolescent

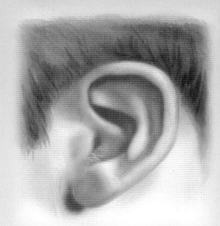

Adult

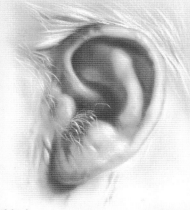

Elderly

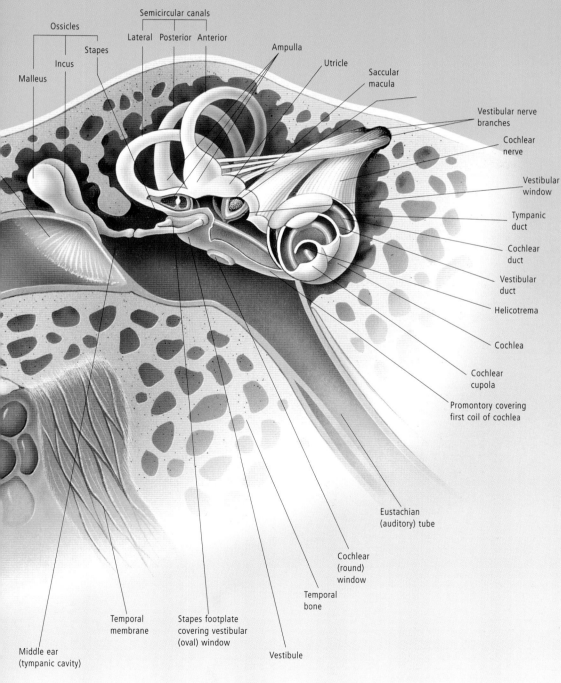

Ossicles
Malleus
Incus
Stapes
Semicircular canals
Lateral
Posterior
Anterior
Ampulla
Utricle
Saccular macula
Vestibular nerve branches
Cochlear nerve
Vestibular window
Tympanic duct
Cochlear duct
Vestibular duct
Helicotrema
Cochlea
Cochlear cupola
Promontory covering first coil of cochlea
Eustachian (auditory) tube
Cochlear (round) window
Temporal bone
Vestibule
Stapes footplate covering vestibular (oval) window
Temporal membrane
Middle ear (tympanic cavity)

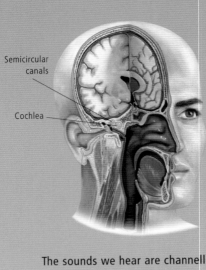

Semicircular canals

Cochlea

The sounds we hear are channelled through the outer ear as sound waves. Our hearing can distinguish a range of sounds and frequencies.

The sound waves hit the eardrum and cause vibrations to be triggered off. These vibrations travel through the middle ear, setting off a chain reaction in the ossicles (the bones known as the hammer, anvil and stirrup), which intensify the vibrations and relay them to the inner ear. The vibrations cause the fluid in the cochlea of the inner ear to pulse, activating the receptor cells located in the organ of Corti.

Ossicles

Semicircular canals

Eardrum

Hearing

Organ of Corti

Tectorial membrane

Inner hair cell

Outer hair cell

Phalangeal cell

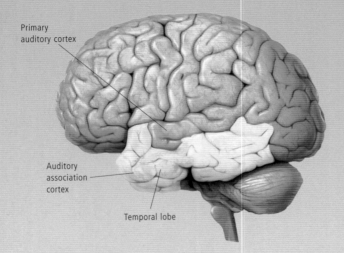

Primary auditory cortex

Auditory association cortex

Temporal lobe

Nerve fibers

Pillar cell

Basilar membrane

The impulses are sent on to the auditory centers of the brain, located in the temporal lobe of the brain, where sounds are interpreted.

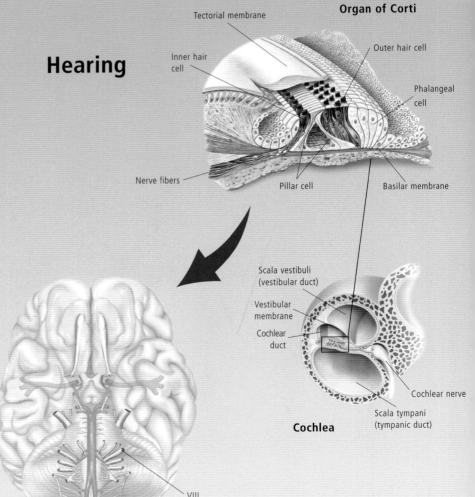

Scala vestibuli (vestibular duct)

Vestibular membrane

Cochlear duct

Cochlear nerve

Scala tympani (tympanic duct)

Cochlea

VIII

The receptor cells (mechanoreceptors) in the organ of Corti send nerve impulses via the vestibulocochlear nerve, the eighth cranial nerve.

Balance mechanism in the ear

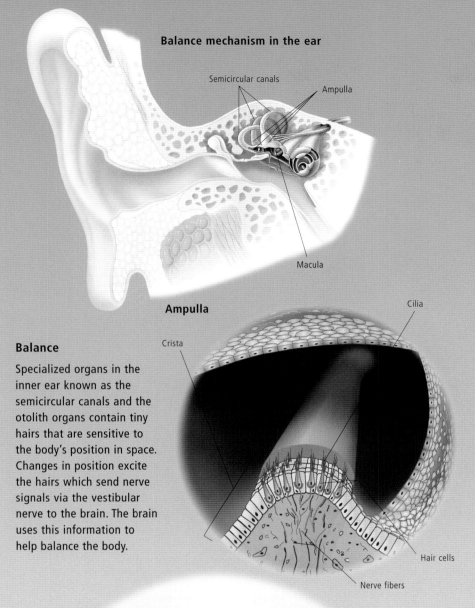

Semicircular canals

Ampulla

Macula

Ampulla

Cilia

Crista

Hair cells

Nerve fibers

Balance

Specialized organs in the inner ear known as the semicircular canals and the otolith organs contain tiny hairs that are sensitive to the body's position in space. Changes in position excite the hairs which send nerve signals via the vestibular nerve to the brain. The brain uses this information to help balance the body.

Macule of saccule

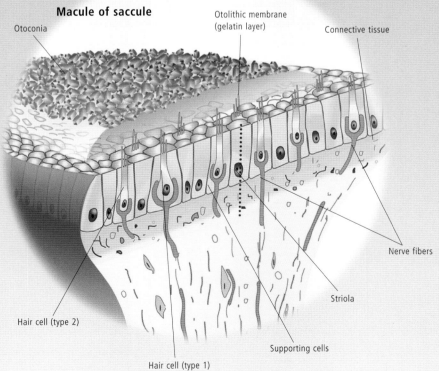

Otoconia

Otolithic membrane (gelatin layer)

Connective tissue

Nerve fibers

Striola

Hair cell (type 2)

Supporting cells

Hair cell (type 1)

Hearing

HOW WE HEAR

Sound waves enter the ear canal and hit the eardrum. The eardrum vibrates and relays vibrations to the ossicles. The ossicles transmit the vibrations to the oval window (a membrane that covers the entrance to the cochlea).

The vibrations pass into the cochlea where fluid activates the tiny hair-like receptor cells (mechanoreceptors) in the organ of Corti. These cells send nerve impulses along the vestibulocochlear nerve to the auditory center in the temporal lobe of the brain where sounds are interpreted.

Balance

Within the ear, movement is registered by the semicircular canals and otolith organs. The otolith organs are hollow sacs containing a gelatinous fluid. Tiny hair cells are anchored to the inner surface of the organs, and above them lie crystals of calcium carbonate, the otoliths. When the head moves, the otoliths change position, activating the hair cells, which send impulses to the brain. The brain then triggers the body's reflex mechanisms to correct the position of the body.

The hair-like nerve cells in the semicircular ducts are also triggered by fluid movement, causing the nerves to send impulses to the brain. The nerves of the semicircular canals and otolith organs transmit impulses via the vestibulocochlear nerve to the brain. While the eyes transmit visual information about the body's position, the semicircular canals and otolith organs, along with the nerves and muscles that control motor co-ordination and movement, maintain the body's balance.

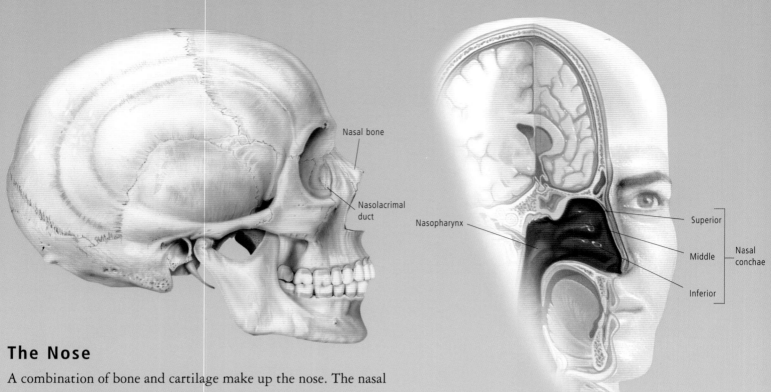

Nasal bone

Nasolacrimal duct

Nasopharynx

Superior

Middle

Inferior

Nasal conchae

The Nose

A combination of bone and cartilage make up the nose. The nasal bone and maxillae comprise the bony component of the external structure, while the nostrils are formed by cartilage. The nasal septum, which divides the two nostrils, is a combination of bone and cartilage. The bones surrounding the nasal cavities are the vomer and parts of the frontal, ethmoid, maxillary and sphenoid bones. The floor of the nasal cavities forms the roof of the palate. The nasal cavities lead to the trachea via the pharynx.

Inhaled air is warmed and moistened in the nose, and any foreign particles are trapped by the hairs lining the nostrils. In the nasal cavities, there are three curved plates, called the conchae; these conchae increase the surface area available to warm and moisten the inhaled air. The nasal cavity is lined with mucous membrane (respiratory mucosa), which changes to olfactory mucosa in the upper region of each nasal cavity. The respiratory mucosa has a covering of minute hairs that capture foreign particles and send them to the nasopharynx. The olfactory mucosa contains special nerve cells, called chemoreceptors, used to determine smell.

The lacrimal gland secretes tears, which keep the eye moist; these tears wash over the eye, draining into the nasal cavities through the nasolacrimal duct, and are then swallowed or blown out.

THE PARANASAL SINUSES

The paranasal sinuses are four pairs of cavities connected to the nose. Passages connect the nose to the sinuses in the frontal, ethmoid, maxillary and sphenoid bones of the skull. The sinuses serve as shock absorbers, lighten the bone in which they occur and add resonance to the voice.

Nose

Part of the respiratory system, the nose is the main passageway for air entering the body. As air travels through the nose and nasal cavities, it is filtered, warmed and moistened in readiness for its intake into the lungs.

Paranasal Sinuses

The cavities in the bones surrounding the nose are air-containing spaces that connect with the nasal cavity. The four paired paranasal sinuses are lined with mucous membrane.

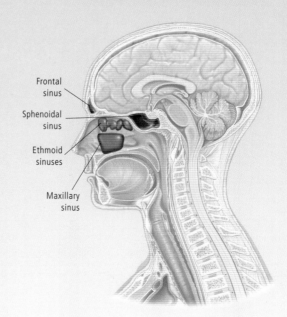

Frontal sinus

Sphenoidal sinus

Ethmoid sinuses

Maxillary sinus

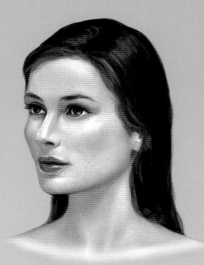

Our nose is part of the respiratory system, and also our special sense organ of smell.

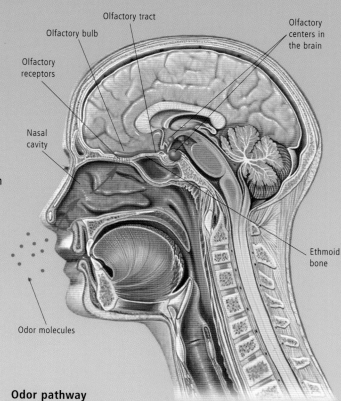

Odor molecules enter the nostrils and pass into the nasal cavity. The nasal cavity has a lining of olfactory mucosa, which contains millions of nerve cells (chemoreceptors). Odors are absorbed by the nerve cells, triggering nerve impulses, which are transmitted to the olfactory bulb, located under the frontal lobe of the brain.

Olfactory tract

Olfactory bulb

Olfactory centers in the brain

Olfactory receptors

Nasal cavity

Ethmoid bone

Odor molecules

Odor pathway

Smell

Closely allied to the sense of taste, smell is one of the body's special senses and our sense of smell can detect a wide range of odors. Because some of the information transmitted by the olfactory nerve is disseminated in parts of the limbic system in the brain, which functions as an area for memory and emotion, so smell can evoke memories of past places and feelings.

Nerve impulses from the olfactory bulb are conducted along the olfactory nerve (cranial nerve I). The impulses are then relayed to the olfactory cortex, the limbic system and the hypothalamus in the brain for identification.

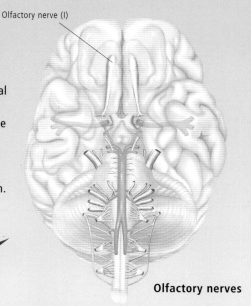

Olfactory nerve (I)

Olfactory nerves

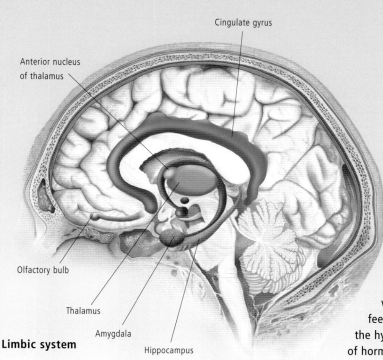

Cingulate gyrus

Anterior nucleus of thalamus

Olfactory bulb

Thalamus

Amygdala

Hippocampus

Limbic system

The limbic system plays an important role in memory and emotion. The olfactory bulb is closely linked to the limbic system, in particular with the hippocampus and amygdala, which is why smells often remind us of past places and feelings. Some smells stimulate the limbic system to activate the hypothalamus and pituitary gland, which triggers the release of hormones associated with appetite and emotional responses.

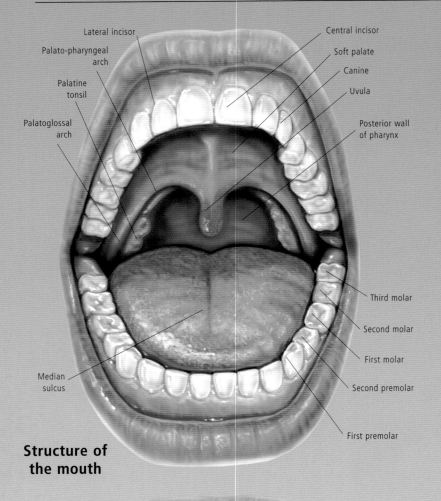

Lateral incisor
Palato-pharyngeal arch
Palatine tonsil
Palatoglossal arch
Median sulcus
Central incisor
Soft palate
Canine
Uvula
Posterior wall of pharynx
Third molar
Second molar
First molar
Second premolar
First premolar

Structure of the mouth

The Mouth

The mouth is the opening between the maxillae and the lower jaw, and is the entrance to the digestive tract. It also connects to the respiratory tract, and is responsible for making sounds, especially speech. It consists of an outer vestibule and an inner true oral cavity. The oral cavity leads to the oropharynx, part of the throat.

The lips form the muscular opening of the mouth and contribute to the formation of words during speech and also help hold food in the mouth. They also help form facial expressions, such as smiling. The sides of the mouth are formed by the muscle tissue of the cheeks. This tissue is covered by skin on the outer surface and by mucous membrane inside the mouth. The cheeks also play an important role in speech and help hold food as it is chewed and then swallowed.

The roof of the mouth comprises the hard palate at the front and the soft palate at the back. Hanging at the back of the soft palate is the droplet-shaped uvula.

The floor of the oral cavity is made up of the tongue and the tissue between the tongue and the teeth.

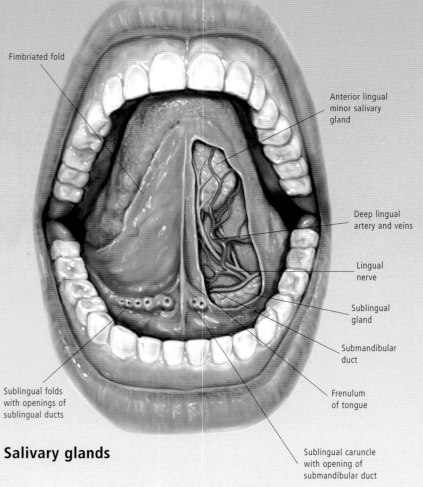

Fimbriated fold
Anterior lingual minor salivary gland
Deep lingual artery and veins
Lingual nerve
Sublingual gland
Submandibular duct
Frenulum of tongue
Sublingual folds with openings of sublingual ducts
Sublingual caruncle with opening of submandibular duct

Salivary glands

Mouth—entrance to body

Involved in a variety of functions, the mouth is the entrance to the digestive tract, plays a role in the respiratory system, and is involved in eating and speaking.

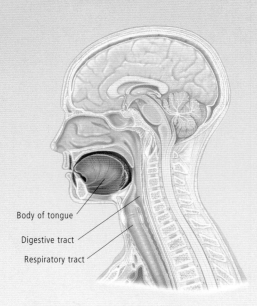

Body of tongue
Digestive tract
Respiratory tract

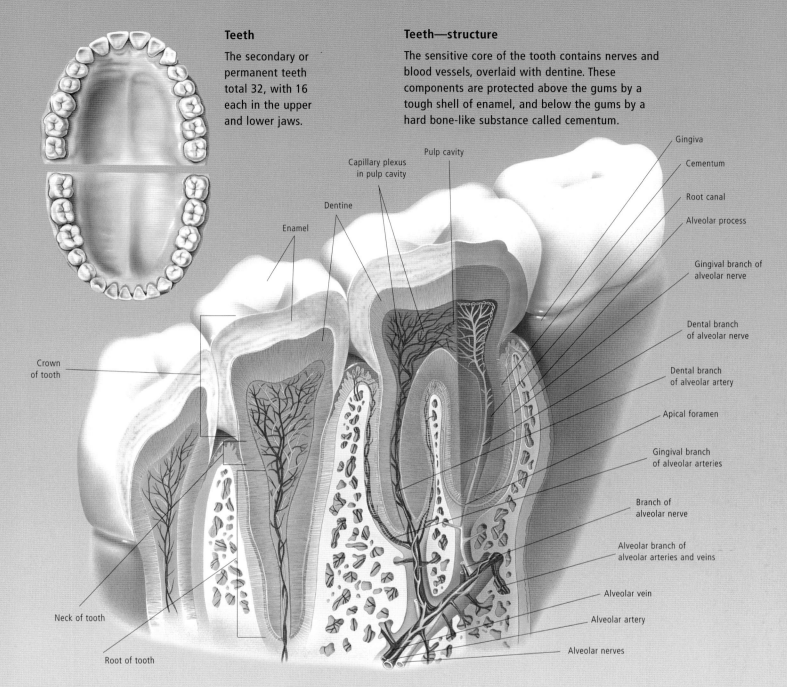

Teeth

The secondary or permanent teeth total 32, with 16 each in the upper and lower jaws.

Teeth—structure

The sensitive core of the tooth contains nerves and blood vessels, overlaid with dentine. These components are protected above the gums by a tough shell of enamel, and below the gums by a hard bone-like substance called cementum.

Pulp cavity

Capillary plexus in pulp cavity

Dentine

Enamel

Crown of tooth

Neck of tooth

Root of tooth

Gingiva

Cementum

Root canal

Alveolar process

Gingival branch of alveolar nerve

Dental branch of alveolar nerve

Dental branch of alveolar artery

Apical foramen

Gingival branch of alveolar arteries

Branch of alveolar nerve

Alveolar branch of alveolar arteries and veins

Alveolar vein

Alveolar artery

Alveolar nerves

The Teeth

Although the teeth appear to be bony in structure, they are, in fact, made up of several layers, with the visible outer layer, or crown of the tooth, being enamel, the hardest substance in the body. Beneath the enamel is dentine, the main component of the teeth, with a softer composition than enamel. The dentine and central pulp are filled with nerves and blood vessels, and below the gum line the dentine is protected by a surrounding layer of tough cementum. Around the cementum is the periodontal ligament, which contains the fibers that anchor the tooth in place.

Two sets of teeth are produced during our lifetime; the first set are the 20 deciduous or primary teeth; these first teeth begin to be replaced from about 7 years of age by the secondary or permanent teeth.

Different types of teeth make up our full set of 32 secondary teeth, including incisors, canines, molars, and premolars. The teeth bite, tear, grind and chew food, processing it to a manageable consistency and size for passage to the digestive tract.

Attached around the neck of the teeth are the gums (gingivae). They extend from inside the lips, around and between the teeth, to the floor of the mouth and the palate. The soft tissue of the gums is kept moist by the salivary glands.

Structures of the Oral Cavity

A variety of structures are contained in the mouth. Apart from the teeth, the mouth contains the tongue, the salivary glands, tonsils and structural features such as the palate, or roof of the mouth. Since our mouth plays a role in eating, speaking and breathing, so the structures within the mouth contribute to these functions.

THE PALATE

Separating the nasal and oral cavities, the palate, or roof of the mouth is made up of two sections, the soft palate and the hard palate. The hard palate section is formed by part of the maxillae, the upper jaw bones, and by the two L-shaped palatine bones. The soft palate behind the hard palate is comprised of muscle tissue, which ends at the back of the mouth. Hanging from its rear edge is the most noticeable feature of the soft palate, the uvula, a teardrop-shaped projection. Whenever we swallow or suck, the soft palate and uvula move up to prevent food from entering the nasal cavity. Both the soft palate and hard palate have a lining of mucous membrane.

THE TONGUE

The tongue is a muscular and sensory organ that is attached to the floor of the mouth. It has a dorsum or upper surface, a base attached to the floor of the mouth, a soft lower surface and a tip.

Since it is required to perform a number of different functions, including chewing, swallowing and speaking, the tongue has a range of movements. The muscles in the tongue are arranged in three directions, allowing the tongue to shorten, narrow, and thin out; these muscles are called the intrinsic tongue muscles. The extrinsic tongue muscles are attached to the jaw, skull, palate and hyoid bones; these muscles allow the tongue to change position, moving it forward, backward, upward and downward. The upper side of the tongue, the dorsum, is covered with three types of papillae (small projections): filiform, fungiform and vallate papillae. Taste buds are found in both the fungiform and vallate papillae; they are also found on the palate, epiglottis and pharynx. The soft underside of the tongue is kept moist by secretions from the salivary glands.

Valleculae   Epiglottis

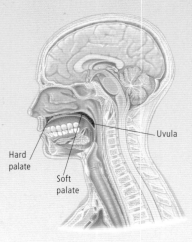

Hard palate

Soft palate

Uvula

Hard Palate

The hard palate extends back from the top teeth, separating the oral and nasal cavities.

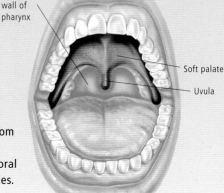

Posterior wall of pharynx

Soft palate

Uvula

Soft Palate

The soft palate is mainly composed of muscle fibers and mucous membrane. The uvula is its most prominent feature.

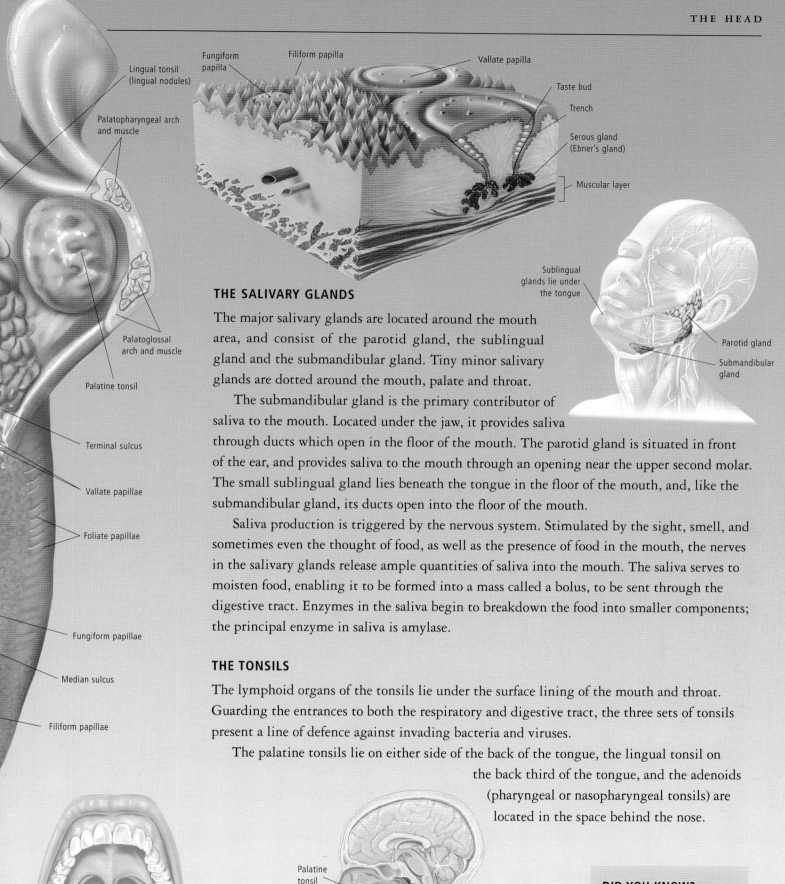

Lingual tonsil (lingual nodules)

Palatopharyngeal arch and muscle

Palatoglossal arch and muscle

Palatine tonsil

Terminal sulcus

Vallate papillae

Foliate papillae

Fungiform papillae

Median sulcus

Filiform papillae

Fungiform papilla

Filiform papilla

Vallate papilla

Taste bud

Trench

Serous gland (Ebner's gland)

Muscular layer

Sublingual glands lie under the tongue

Parotid gland

Submandibular gland

THE SALIVARY GLANDS

The major salivary glands are located around the mouth area, and consist of the parotid gland, the sublingual gland and the submandibular gland. Tiny minor salivary glands are dotted around the mouth, palate and throat.

The submandibular gland is the primary contributor of saliva to the mouth. Located under the jaw, it provides saliva through ducts which open in the floor of the mouth. The parotid gland is situated in front of the ear, and provides saliva to the mouth through an opening near the upper second molar. The small sublingual gland lies beneath the tongue in the floor of the mouth, and, like the submandibular gland, its ducts open into the floor of the mouth.

Saliva production is triggered by the nervous system. Stimulated by the sight, smell, and sometimes even the thought of food, as well as the presence of food in the mouth, the nerves in the salivary glands release ample quantities of saliva into the mouth. The saliva serves to moisten food, enabling it to be formed into a mass called a bolus, to be sent through the digestive tract. Enzymes in the saliva begin to breakdown the food into smaller components; the principal enzyme in saliva is amylase.

THE TONSILS

The lymphoid organs of the tonsils lie under the surface lining of the mouth and throat. Guarding the entrances to both the respiratory and digestive tract, the three sets of tonsils present a line of defence against invading bacteria and viruses.

The palatine tonsils lie on either side of the back of the tongue, the lingual tonsil on the back third of the tongue, and the adenoids (pharyngeal or nasopharyngeal tonsils) are located in the space behind the nose.

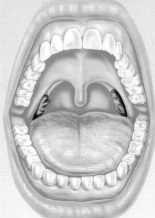

Palatine tonsil

Lingual tonsil

Pharyngeal tonsil

Tonsils

The lymphoid organs of the tonsils protect the entrances to the respiratory and digestive systems.

DID YOU KNOW?

Relative to its size, the tongue is the strongest muscle in the body and is the only muscle in the body which is attached at one end only.

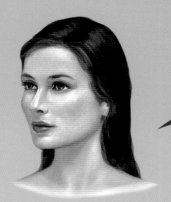

Our sense of taste, also known as gustation, is aided by our sense of smell. In fact, 80 percent of the sensation of taste is due to smell.

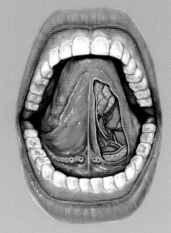

Saliva released by the salivary glands mixes with food. When the mixture comes into contact with the taste buds, it activates the taste buds, which then send nerve impulses to the brain.

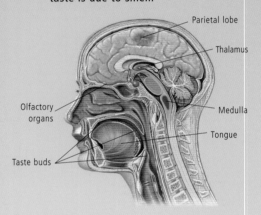

Parietal lobe
Thalamus
Olfactory organs
Medulla
Tongue
Taste buds

Nerve impulses are sent via the cranial nerves to the area of the parietal lobe responsible for taste identification.

Taste

One of the five special senses, our sense of taste is closely allied to, and enhanced by, our sense of smell. The taste buds are activated when food reacts with saliva. The taste buds can detect four basic tastes: sweet, sour, salty and bitter.

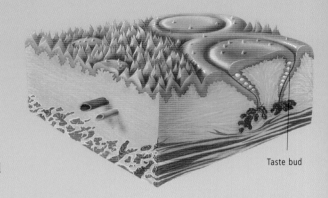

Taste bud

The majority of the taste buds are found in the fungiform and vallate papillae of the tongue, with a small amount of taste buds scattered throughout the back of the mouth and the throat.

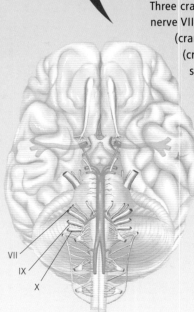

VII
IX
X

Three cranial nerves, the facial nerve (cranial nerve VII), the glossopharyngeal nerve (cranial nerve IX) and the vagus nerve (cranial nerve X) are involved in our sense of taste. Taste buds on the front part of the tongue activate the facial nerve; taste buds on the back of the tongue activate the glossopharyngeal nerve; and the vagus nerve is triggered by the taste buds found in the throat. Nerve impulses are initiated by the activated taste buds and are transmitted along these three nerves to the brain.

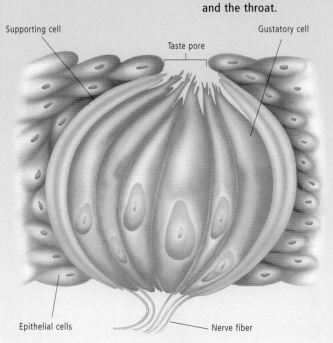

Supporting cell
Gustatory cell
Taste pore
Epithelial cells
Nerve fiber

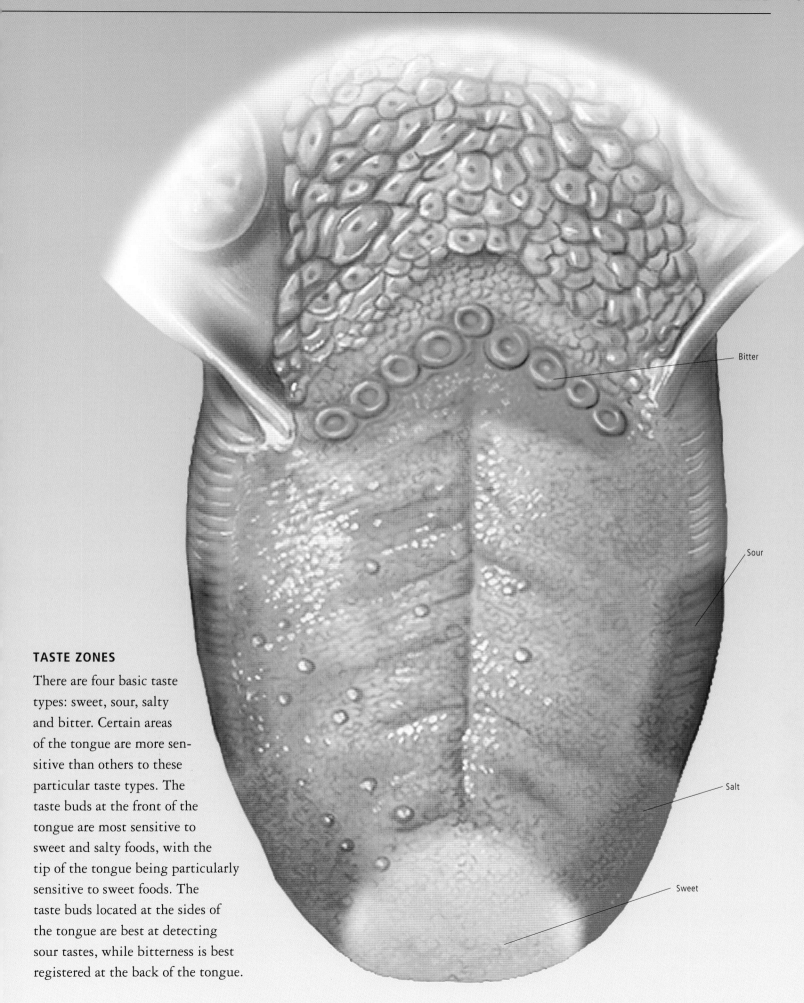

Bitter

Sour

Salt

Sweet

TASTE ZONES

There are four basic taste
types: sweet, sour, salty
and bitter. Certain areas
of the tongue are more sensitive than others to these
particular taste types. The
taste buds at the front of the
tongue are most sensitive to
sweet and salty foods, with the
tip of the tongue being particularly
sensitive to sweet foods. The
taste buds located at the sides of
the tongue are best at detecting
sour tastes, while bitterness is best
registered at the back of the tongue.

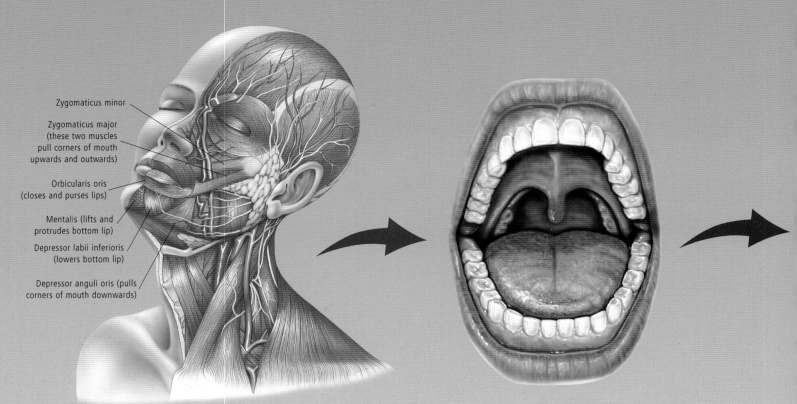

Zygomaticus minor

Zygomaticus major
(these two muscles
pull corners of mouth
upwards and outwards)

Orbicularis oris
(closes and purses lips)

Mentalis (lifts and
protrudes bottom lip)

Depressor labii inferioris
(lowers bottom lip)

Depressor anguli oris (pulls
corners of mouth downwards)

Some sounds generated require the tongue to come into contact with other structures in the mouth to create speech. The tongue meets with the teeth to make the sound "t", meets with the soft palate to make the sound "g", and meets with the hard palate to make "n".

Speech

Speech production involves the coordinated involvement of various organs and structures in the body. Firstly, air is expelled from the lungs. Then the air passes through the adjusted vocal cords, altering the air flow and creating vibrations which produce sound. Synchronized movements of the muscles of the mouth and tongue, aided by the soft palate, tongue and lips, modify the sound generated by the vocal cords, to produce speech.

DID YOU KNOW?

The coordinated movements of 72 muscles are involved in the production of speech.

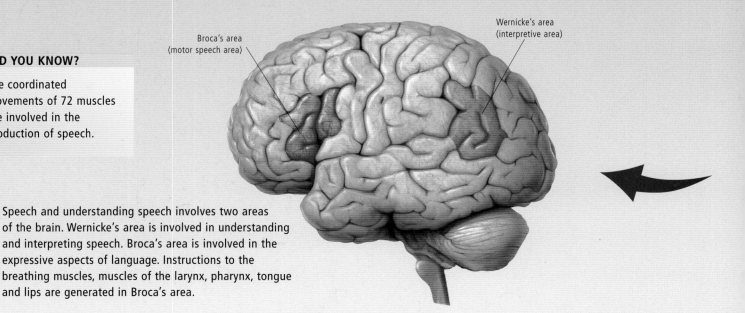

Broca's area
(motor speech area)

Wernicke's area
(interpretive area)

Speech and understanding speech involves two areas of the brain. Wernicke's area is involved in understanding and interpreting speech. Broca's area is involved in the expressive aspects of language. Instructions to the breathing muscles, muscles of the larynx, pharynx, tongue and lips are generated in Broca's area.

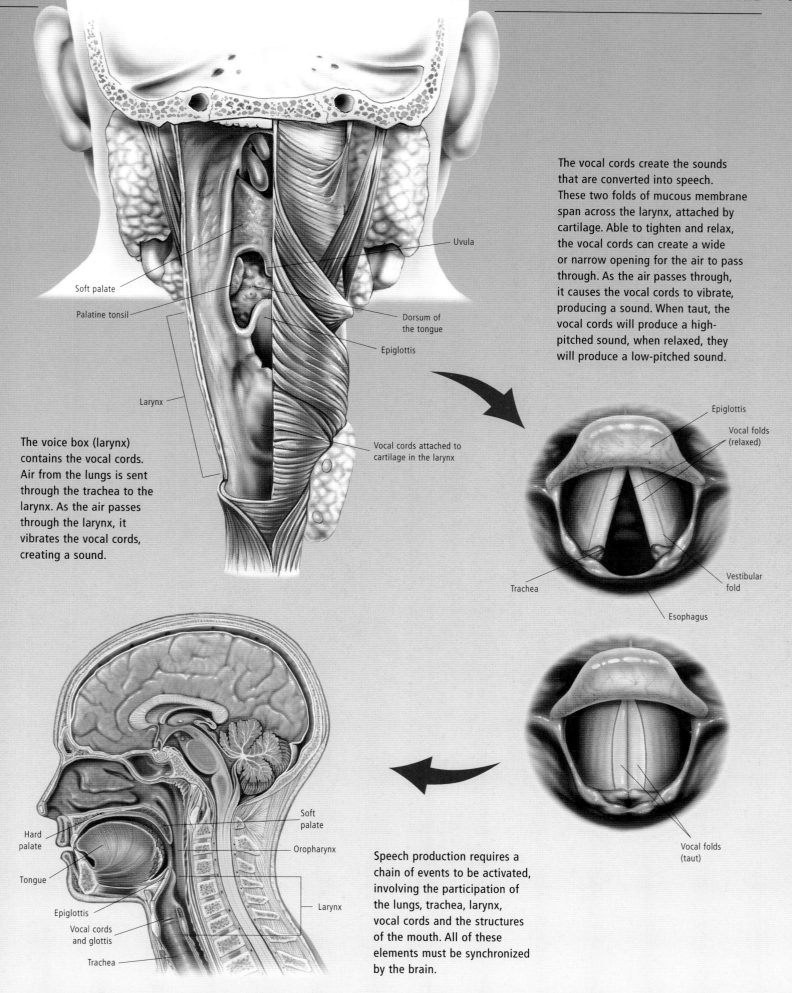

Uvula

Soft palate

Palatine tonsil

Dorsum of
the tongue

Epiglottis

Larynx

The vocal cords create the sounds
that are converted into speech.
These two folds of mucous membrane
span across the larynx, attached by
cartilage. Able to tighten and relax,
the vocal cords can create a wide
or narrow opening for the air to pass
through. As the air passes through,
it causes the vocal cords to vibrate,
producing a sound. When taut, the
vocal cords will produce a high-
pitched sound, when relaxed, they
will produce a low-pitched sound.

The voice box (larynx)
contains the vocal cords.
Air from the lungs is sent
through the trachea to the
larynx. As the air passes
through the larynx, it
vibrates the vocal cords,
creating a sound.

Vocal cords attached to
cartilage in the larynx

Epiglottis

Vocal folds
(relaxed)

Vestibular
fold

Trachea

Esophagus

Vocal folds
(taut)

Soft
palate

Oropharynx

Larynx

Hard
palate

Tongue

Epiglottis

Vocal cords
and glottis

Trachea

Speech production requires a
chain of events to be activated,
involving the participation of
the lungs, trachea, larynx,
vocal cords and the structures
of the mouth. All of these
elements must be synchronized
by the brain.

The Neck

The Structure and Function of the Neck

The neck supports and provides mobility for the head. The spinal cord, protected by the vertebrae, major blood vessels and nerves of the brain and face, and passageways for food and air can all be found in the neck.

The neck can be divided into two major columns. The back column comprises the seven vertebrae of the neck (cervical vertebrae) and their supporting musculature. The front column of the neck contains the pharynx, larynx, trachea, and esophagus—the passageways to the lungs and stomach.

The thyroid gland is attached to the front and sides of the trachea and larynx, and covering these structures at the front are thin strap-like muscles and skin. Behind, and sometimes embedded within, the thyroid are the tiny parathyroid glands.

Neck

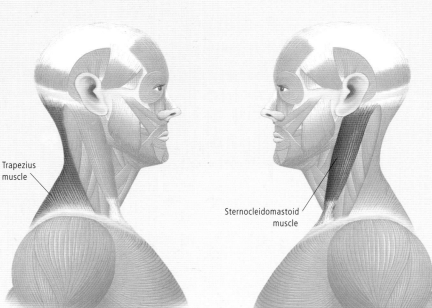

Trapezius muscle

Sternocleidomastoid muscle

Neck muscles

The trapezius muscle and sternocleidomastoid muscle are two of the largest muscles in the neck.

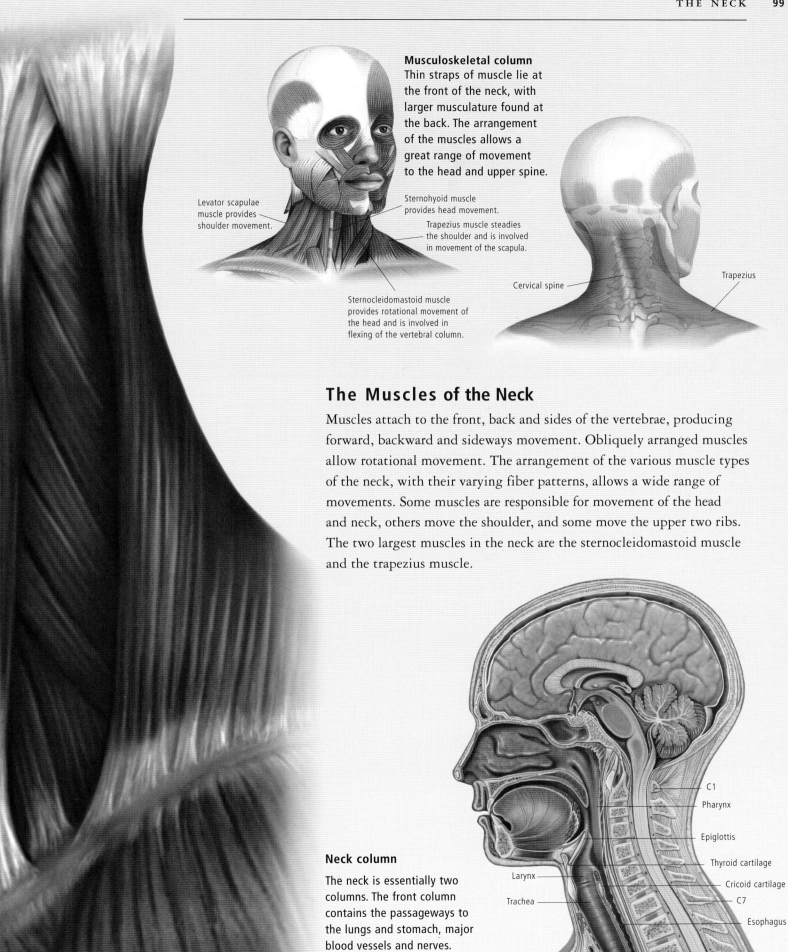

Musculoskeletal column
Thin straps of muscle lie at the front of the neck, with larger musculature found at the back. The arrangement of the muscles allows a great range of movement to the head and upper spine.

Levator scapulae muscle provides shoulder movement.

Sternohyoid muscle provides head movement.

Trapezius muscle steadies the shoulder and is involved in movement of the scapula.

Sternocleidomastoid muscle provides rotational movement of the head and is involved in flexing of the vertebral column.

Cervical spine

Trapezius

The Muscles of the Neck

Muscles attach to the front, back and sides of the vertebrae, producing forward, backward and sideways movement. Obliquely arranged muscles allow rotational movement. The arrangement of the various muscle types of the neck, with their varying fiber patterns, allows a wide range of movements. Some muscles are responsible for movement of the head and neck, others move the shoulder, and some move the upper two ribs. The two largest muscles in the neck are the sternocleidomastoid muscle and the trapezius muscle.

Neck column

The neck is essentially two columns. The front column contains the passageways to the lungs and stomach, major blood vessels and nerves. The back column contains the cervical vertebrae of the spine and the spinal cord.

C1

Pharynx

Epiglottis

Thyroid cartilage

Larynx

Cricoid cartilage

Trachea

C7

Esophagus

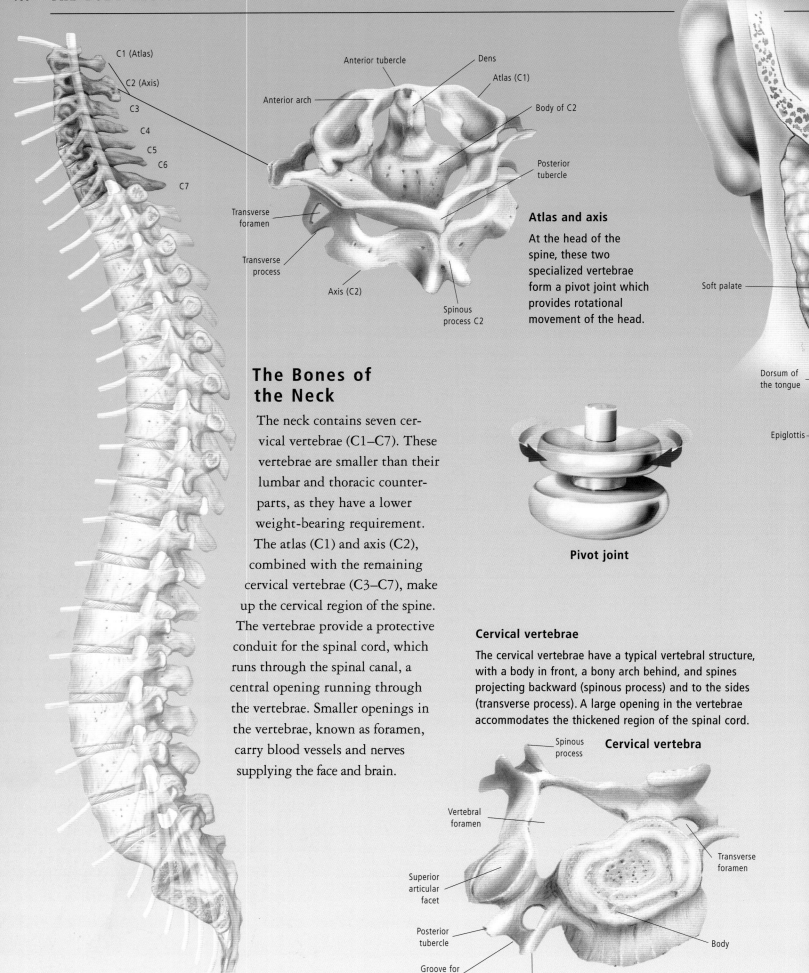

C1 (Atlas)

C2 (Axis)

C3

C4

C5

C6

C7

Anterior tubercle

Dens

Atlas (C1)

Anterior arch

Body of C2

Posterior tubercle

Transverse foramen

Transverse process

Axis (C2)

Spinous process C2

Soft palate

Dorsum of the tongue

Epiglottis

Atlas and axis

At the head of the spine, these two specialized vertebrae form a pivot joint which provides rotational movement of the head.

The Bones of the Neck

The neck contains seven cervical vertebrae (C1–C7). These vertebrae are smaller than their lumbar and thoracic counterparts, as they have a lower weight-bearing requirement. The atlas (C1) and axis (C2), combined with the remaining cervical vertebrae (C3–C7), make up the cervical region of the spine. The vertebrae provide a protective conduit for the spinal cord, which runs through the spinal canal, a central opening running through the vertebrae. Smaller openings in the vertebrae, known as foramen, carry blood vessels and nerves supplying the face and brain.

Pivot joint

Cervical vertebrae

The cervical vertebrae have a typical vertebral structure, with a body in front, a bony arch behind, and spines projecting backward (spinous process) and to the sides (transverse process). A large opening in the vertebrae accommodates the thickened region of the spinal cord.

Spinous process

Cervical vertebra

Vertebral foramen

Transverse foramen

Superior articular facet

Posterior tubercle

Groove for spinal nerve

Anterior tubercle

Body

The Throat

The throat is located in the front section of the neck. Within it are the fauces, the opening that leads from the back of the mouth into the pharynx, and the pharynx itself, the cavity that connects the mouth, nose and larynx, which is situated behind the arch at the back of the mouth. One of the most important regions of the throat, the pharynx contains structures used in the processes of breathing, speaking and swallowing. The pharynx is comprised of three sections, the nasopharynx, the oropharynx and the laryngopharynx, and is a common passageway for both the respiratory and digestive processes.

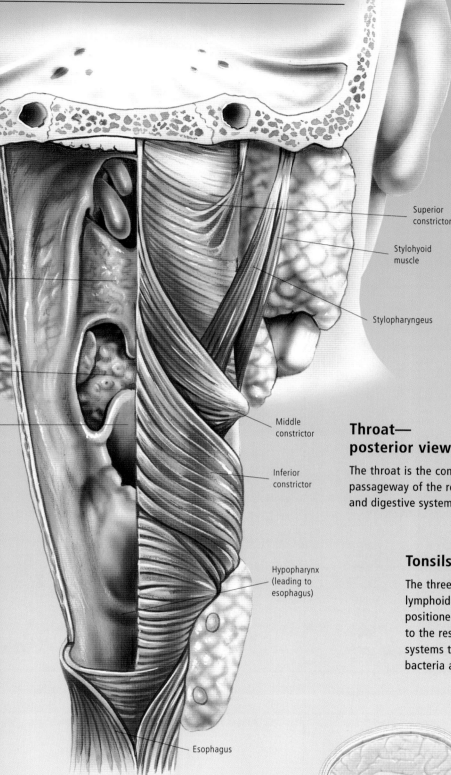

Superior constrictor

Stylohyoid muscle

Stylopharyngeus

Middle constrictor

Inferior constrictor

Hypopharynx (leading to esophagus)

Esophagus

Throat— posterior view

The throat is the common passageway of the respiratory and digestive systems.

Tonsils

The three sets of tonsils are lymphoid organs, strategically positioned around the entrances to the respiratory and digestive systems to protect the body from bacteria and viruses.

Pharyngeal tonsil

Palatine tonsil

Lingual tonsil

Nasopharynx

The nasopharynx is found immediately beneath the base of the skull and behind the nose, and contains the adenoids (pharyngeal tonsils), and the openings to the eustachian tubes which connect to the middle ear.

Oropharynx

Lying at the back of the mouth, the oropharynx contains the tonsils and the back of the tongue, and provides a passage for air, water and food.

Pharynx

The pharynx comprises three regions: the nasopharynx, the oropharynx and the laryngopharynx.

Laryngopharynx

The laryngopharynx contains the epiglottis and voice box (larynx), which lead to the lungs. Food passes through the laryngopharynx on its way to the esophagus and stomach.

The Larynx

The voice box (larynx) is the part of the throat that leads from the pharynx to the windpipe (trachea) and lungs. The larynx serves two main functions: to protect the airway to the lungs from inhalation of food and liquids; and to produce a source of air vibration for the voice. The larynx is composed of nine cartilages which provide strength for the airway and attachments for the associated muscles, ligaments and membranes.

Lying immediately behind and below the tongue is the epiglottis, a flap of elastic cartilage that can be controlled by muscles to close off the entrance of the larynx during swallowing, thus preventing food and liquids from entering the airways.

The larynx contains the vocal cords, two pairs of mucous membranes, controlled by the laryngeal muscles. During breathing, the vocal cords are separated by a moderate opening to allow the passage of air. For speech production, this opening narrows and the cords are tensed. Air forced through the cords causes them to vibrate, producing sound waves.

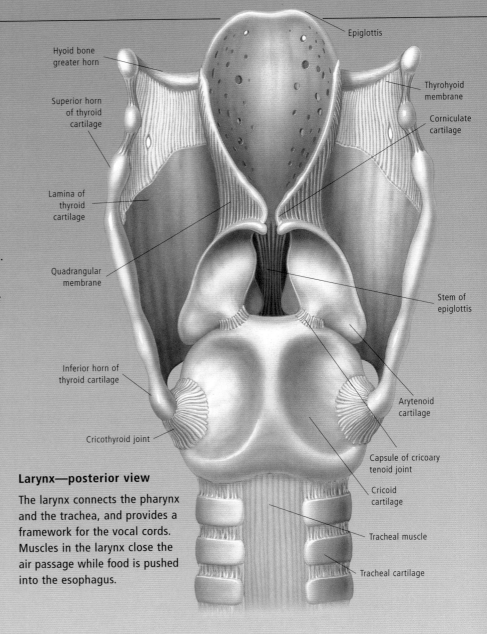

Larynx—posterior view

The larynx connects the pharynx and the trachea, and provides a framework for the vocal cords. Muscles in the larynx close the air passage while food is pushed into the esophagus.

Epiglottis—closed

When swallowing, muscles of the pharynx lift the larynx and close the epiglottis, effectively sealing off the trachea and preventing food and liquids from entering the airways.

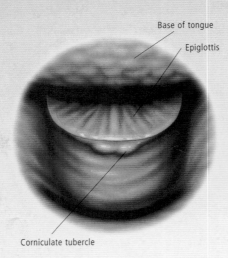

Vocal cords—closed

When the vocal cords are tensed, air passing through the small gap causes the cords to vibrate, producing sound waves.

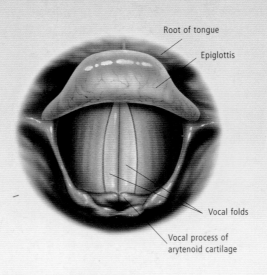

Vocal cords—open

When the vocal cords are relaxed, they have a moderate opening which allows air to be inhaled and exhaled without vibration, thus no sound is produced.

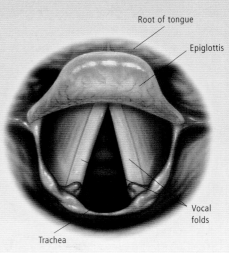

The Esophagus

Passing through the neck and chest on its way to the stomach, the esophagus is a muscular tube that allows food to be transported from the throat (pharynx) to the stomach. It can expand and contract at its upper and lower ends using circular muscles called sphincters. The upper sphincter relaxes to accept food from the pharynx, and the food is moved by muscular contractions to the lower sphincter, which relaxes to allow food to enter the stomach.

The Trachea

The windpipe (trachea) is a fibro-elastic and muscular tube for the passage of air, beginning at the lower end of the voice box (larynx), and ending at the two main bronchi.

The trachea is reinforced with C-shaped cartilages which prevent collapse of the airway, and elastic fibers which allow it to stretch and recoil with the movements of the larynx (which is used in swallowing and speech) and diaphragm (used in breathing). The back of the trachea is flat, with the ends of the cartilage bridged by the trachealis muscle; the contractions of this muscle reduce the diameter of the airway. Lying against this surface is the esophagus, which expands into the gap in the cartilage when food is swallowed.

Dust particles in the trachea become trapped in the mucous membrane lining, and tiny hair-like projections on the membrane (cilia), move the dust-laden mucus toward the throat where it may be swallowed or expelled.

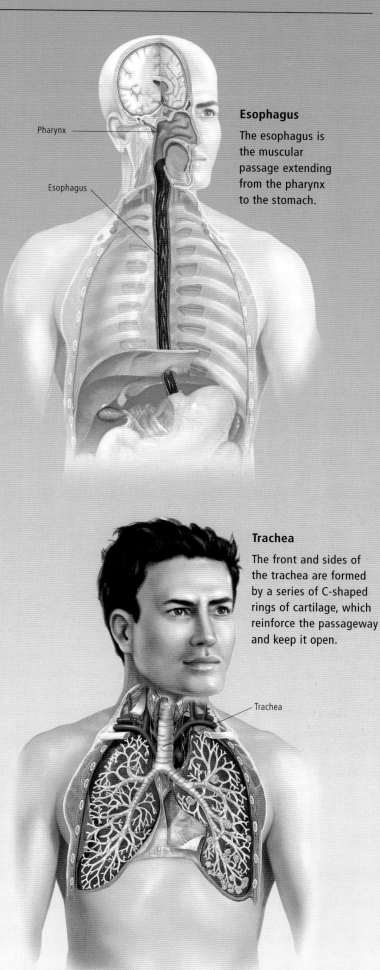

Esophagus

The esophagus is the muscular passage extending from the pharynx to the stomach.

Pharynx

Esophagus

Trachea

The front and sides of the trachea are formed by a series of C-shaped rings of cartilage, which reinforce the passageway and keep it open.

Trachea

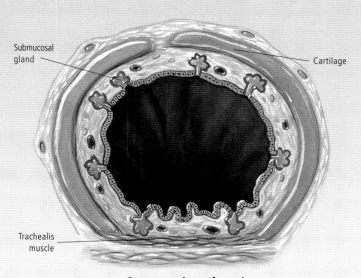

Submucosal gland

Cartilage

Trachealis muscle

Cross-section of trachea

Major Glands and Blood Vessels of the Neck

THYROID AND PARATHYROID GLANDS

The thyroid and parathyroid glands are members of the endocrine system. The largest of the endocrine glands, the thyroid is located directly below and in front of the voice box (larynx), and consists of two lobes joined at the midline by a narrow bridge. The thyroid gland secretes thyroid hormone directly into the bloodstream and body cavities, and its prime function is to control the metabolic rate of body tissues.

Lying just behind the thyroid gland are the four (or occasionally three) parathyroid glands. These tiny glands produce parathyroid hormone, which is involved in the control of calcium and phosphate concentrations in the blood. When blood calcium levels fall, parathyroid hormone stimulates the release of calcium from the bones. If blood calcium levels are too high, the thyroid gland releases a hormone called calcitonin to lower the levels. The secretion of both thyroid hormone and parathyroid hormone is initiated by the pituitary gland, which monitors the body's hormonal levels. The pituitary gland is the control center of the endocrine system; it responds to fluctuations in the body's hormonal requirements by activating the endocrine organs, including the thyroid and parathyroid glands, when needed.

DID YOU KNOW?

The larynx grows rapidly in males during puberty, creating an increased prominence of the ridge and notch of the thyroid cartilage—the Adam's apple. The elongated vocal ligaments vibrate with a lower frequency, resulting in a deeper voice.

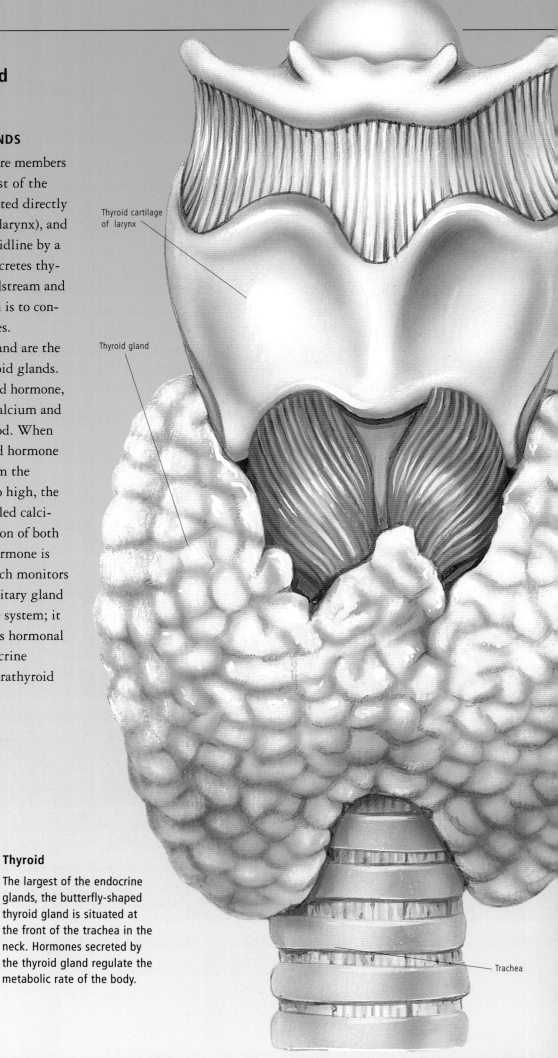

Thyroid cartilage of larynx

Thyroid gland

Trachea

Thyroid

The largest of the endocrine glands, the butterfly-shaped thyroid gland is situated at the front of the trachea in the neck. Hormones secreted by the thyroid gland regulate the metabolic rate of the body.

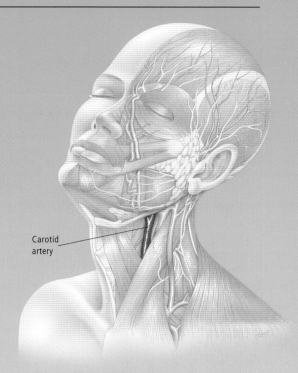

Jugular vein

The jugular veins are responsible for carrying blood from the head and neck back to the heart. Valves in the veins prevent the backflow of blood into the brain.

Internal jugular vein

External jugular vein

Carotid artery

Carotid artery

The carotid arteries are the major arteries running through the neck, supplying blood to the neck, head and brain.

THE BLOOD VESSELS OF THE NECK

On either side of the neck are the large blood vessels that supply the neck, head and brain with blood. Found in the lower part of the neck, bound within a sheath of connective tissue, is the common carotid artery, that divides into internal and external carotid arteries at about the level of the Adam's apple. The internal carotid artery supplies blood from the aorta to the brain, while the external carotid artery supplies the face and neck.

Two jugular veins (an internal jugular vein and an external jugular vein) on either side of the neck drain blood from the brain, face and neck. These veins have a valve at their lower ends to prevent blood flowing back toward the brain.

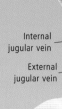

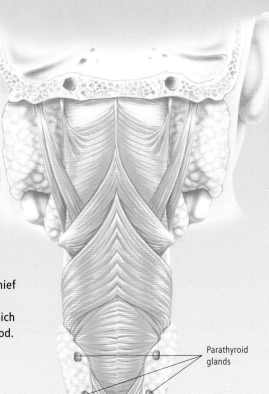

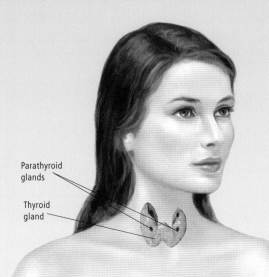

Parathyroid glands

Thyroid gland

Parathyroid glands

These tiny, pea-sized glands are found behind the thyroid gland. Within the fibrous capsule of the gland, there are two cell types, chief and oxyphil cells. The chief cells produce parathyroid hormone which controls calcium levels in the blood.

Parathyroid glands

The Trunk

Spine

The spine is a chain of bones known as vertebrae. Between each vertebra is a cushioning pad of cartilage known as the intervertebral disk. Although each vertebra has limited movement, the sum of these individual movements results in great mobility.

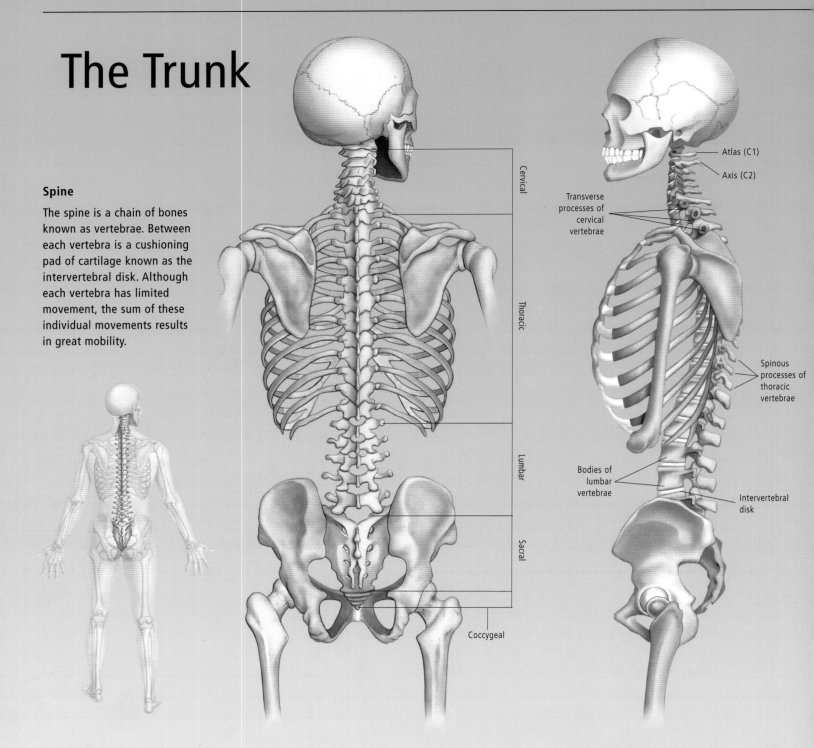

Cervical

Thoracic

Lumbar

Sacral

Coccygeal

Atlas (C1)

Axis (C2)

Transverse processes of cervical vertebrae

Spinous processes of thoracic vertebrae

Bodies of lumbar vertebrae

Intervertebral disk

The Back

The back is the part of the human body from just below the neck to the lower end of the spine, just above the buttocks. The bones of the spine are known as the vertebrae and are joined to each other by flexible disks (intervertebral disks) and joints (facet joints) to form the vertebral column. The vertebral column protects the spinal cord and spinal nerves, supports the weight of the body and head, and anchors the rib cage. It plays a major role in

movement and posture, and provides an attachment point for muscles of the trunk, arms and legs. Strong back and abdominal muscles support the vertebral column, and are involved in the breathing process.

Composed of nerve tissue, the spinal cord runs through a canal formed by the vertebrae, and nerves branch off through spaces between the vertebrae. The arteries that supply blood to the brain stem travel up the cervical vertebrae.

Surface muscles

Deep muscles

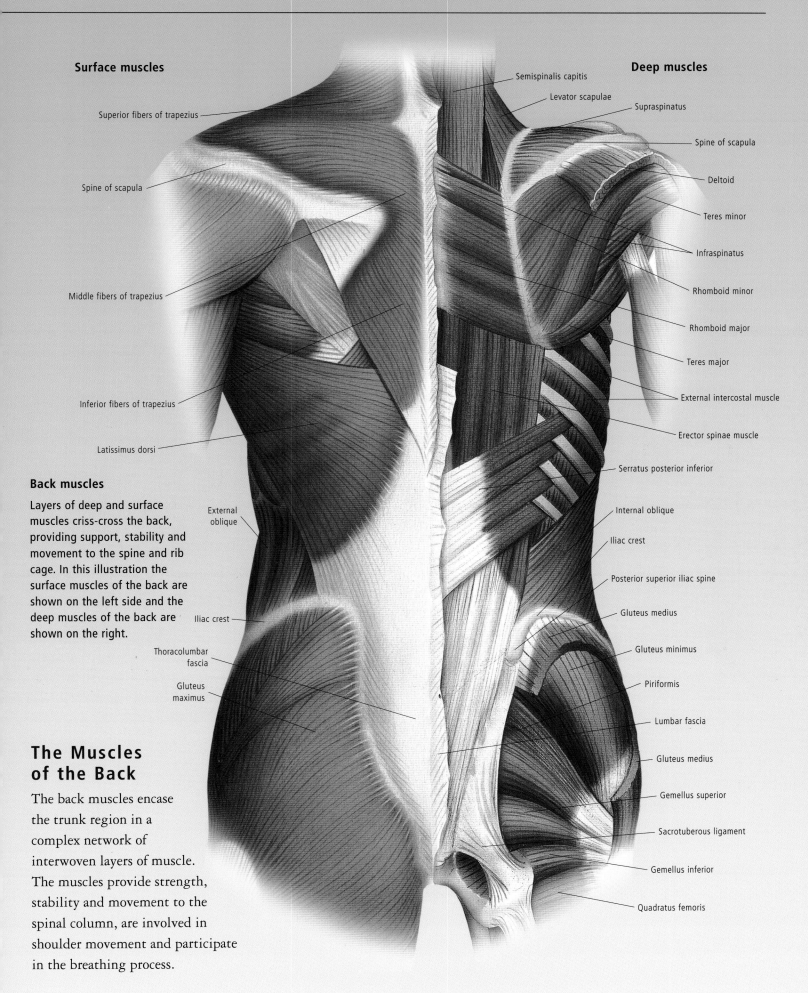

Semispinalis capitis

Levator scapulae

Supraspinatus

Superior fibers of trapezius

Spine of scapula

Spine of scapula

Deltoid

Teres minor

Infraspinatus

Middle fibers of trapezius

Rhomboid minor

Rhomboid major

Teres major

External intercostal muscle

Inferior fibers of trapezius

Erector spinae muscle

Latissimus dorsi

Serratus posterior inferior

Back muscles

Layers of deep and surface muscles criss-cross the back, providing support, stability and movement to the spine and rib cage. In this illustration the surface muscles of the back are shown on the left side and the deep muscles of the back are shown on the right.

External oblique

Internal oblique

Iliac crest

Posterior superior iliac spine

Iliac crest

Gluteus medius

Gluteus minimus

Thoracolumbar fascia

Piriformis

Gluteus maximus

Lumbar fascia

The Muscles of the Back

The back muscles encase the trunk region in a complex network of interwoven layers of muscle. The muscles provide strength, stability and movement to the spinal column, are involved in shoulder movement and participate in the breathing process.

Gluteus medius

Gemellus superior

Sacrotuberous ligament

Gemellus inferior

Quadratus femoris

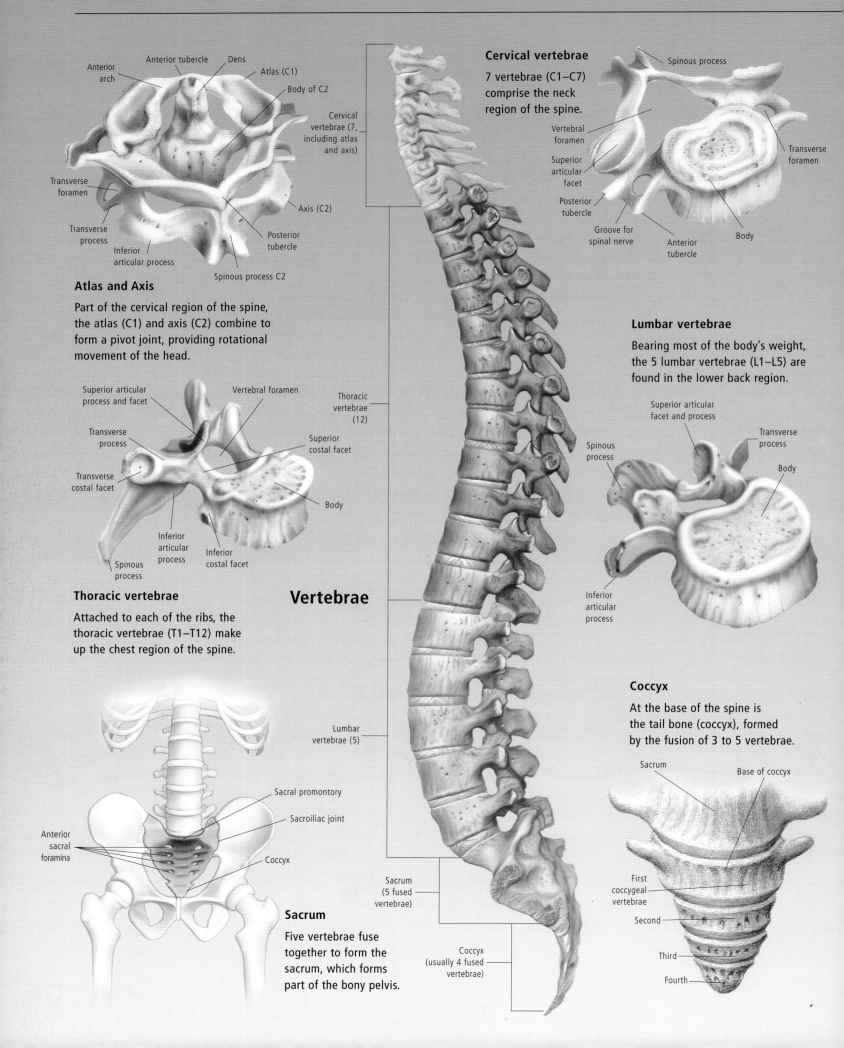

Anterior arch
Anterior tubercle
Dens
Atlas (C1)
Body of C2
Cervical vertebrae (7, including atlas and axis)
Transverse foramen
Transverse process
Inferior articular process
Posterior tubercle
Spinous process C2
Axis (C2)

Atlas and Axis

Part of the cervical region of the spine, the atlas (C1) and axis (C2) combine to form a pivot joint, providing rotational movement of the head.

Superior articular process and facet
Transverse process
Transverse costal facet
Vertebral foramen
Superior costal facet
Body
Inferior articular process
Inferior costal facet
Spinous process

Thoracic vertebrae

Attached to each of the ribs, the thoracic vertebrae (T1–T12) make up the chest region of the spine.

Vertebrae

Thoracic vertebrae (12)

Lumbar vertebrae (5)

Sacrum (5 fused vertebrae)

Coccyx (usually 4 fused vertebrae)

Cervical vertebrae

7 vertebrae (C1–C7) comprise the neck region of the spine.

Spinous process
Vertebral foramen
Superior articular facet
Posterior tubercle
Groove for spinal nerve
Anterior tubercle
Transverse foramen
Body

Lumbar vertebrae

Bearing most of the body's weight, the 5 lumbar vertebrae (L1–L5) are found in the lower back region.

Superior articular facet and process
Spinous process
Transverse process
Body
Inferior articular process

Coccyx

At the base of the spine is the tail bone (coccyx), formed by the fusion of 3 to 5 vertebrae.

Sacrum
Base of coccyx
First coccygeal vertebrae
Second
Third
Fourth

Sacral promontory
Sacroiliac joint
Anterior sacral foramina
Coccyx

Sacrum

Five vertebrae fuse together to form the sacrum, which forms part of the bony pelvis.

The Spine and Vertebrae

Forming the central axis of the skeleton, the spine, or vertebral column, extends down the back, from the base of the skull to the pelvis. This structure of vertebrae, stacked one upon another and each separated by an intervertebral disk, is strong enough to support the weight of the upper body and head, yet flexible enough to allow movements such as bending and twisting.

The spine is comprised of 24 separate vertebrae: 7 cervical (C1–7) in the neck, 12 thoracic (T1–12) in the chest, and 5 lumbar (L1–5) in the small of the back; the sacrum and the coccyx. The neck region commences with the atlas and axis (C1 and C2), a specially designed pivot joint that provides the head with rotational movement. This joint, combined with the remaining cervical vertebrae, C3–C7, completes the cervical region of the spine, one of the most flexible of the spinal regions.

The thoracic vertebrae, T1–T12, are found in the chest region. The thoracic vertebrae are attached to the ribs, which results in less flexibility than both the cervical and lumbar regions.

Responsible for bearing most of the body's weight, the largest of the vertebrae, the lumbar vertebrae (L1–L5), are found in the lower back region. Despite their load-bearing requirements, the lumbar vertebrae still allow great flexibility and mobility.

Each vertebra in the spine forms three separate joints with the vertebra above and below—a pair of facet joints and a single anterior intervertebral joint. The direction and orientation of the joint surfaces determines the type of movements that are allowed in different regions of the vertebral column.

The sacrum begins life as 5 separate vertebrae. During early development, these 5 vertebrae fuse together to form one unit, the sacrum. Forming part of the pelvis, the sacrum is joined to the hips at the sacroiliac joint.

Attached to the sacrum, the coccyx, or tail bone, is the lowest part of the spine. Originating as 3 to 5 vertebrae, these bones fuse together during development to form the coccyx.

INTERVERTEBRAL DISK

The intervertebral disks lie between each of the vertebrae. They are designed to bear weight, and to act as a cushion between the individual vertebrae when the spine moves. The pliable cartilaginous disks unite the stacked column of individual vertebrae, transforming them into a strong yet supple complete unit, the back bone.

The Pectoral Girdle

The bones and muscles of the shoulder girdle (pectoral girdle) support the shoulder joint. One set of muscles attaches the shoulder girdle to the trunk of the body, while another set of muscles joins the shoulder girdle to the humerus bone of the arm.

The Trapezius Muscle

The trapezius muscle is found lying under the skin of the back of the neck and upper part of the back of the chest. The trapezius muscle spans between the vertebrae of the thorax and neck and the occipital bone of the cranium across to the collar bone (clavicle) and shoulder blade (scapula). The main function of the trapezius muscle is to steady the shoulder; it also assists in movements of the scapula.

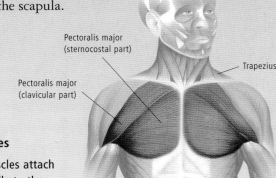

Pectoralis major (sternocostal part)

Trapezius

Pectoralis major (clavicular part)

Pectoral muscles

The pectoral muscles attach the shoulder girdle to the trunk of the body, and stabilize the shoulder joint.

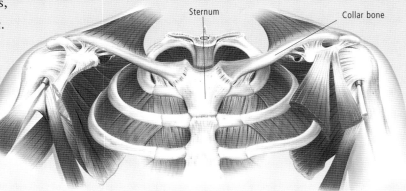

Sternum

Collar bone

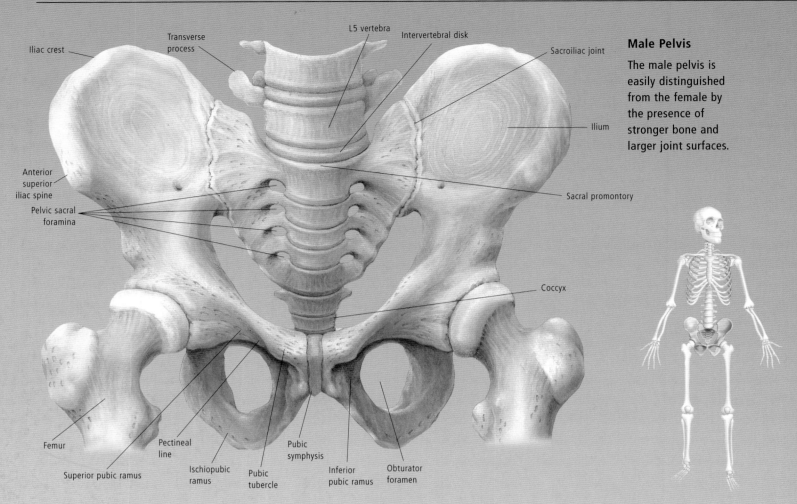

Iliac crest

Transverse process

L5 vertebra

Intervertebral disk

Sacroiliac joint

Ilium

Anterior superior iliac spine

Pelvic sacral foramina

Sacral promontory

Coccyx

Femur

Pectineal line

Ischiopubic ramus

Pubic tubercle

Pubic symphysis

Inferior pubic ramus

Obturator foramen

Superior pubic ramus

Male Pelvis

The male pelvis is easily distinguished from the female by the presence of stronger bone and larger joint surfaces.

The Pelvis

Protecting the lower abdominal organs, the pelvis can refer to a number of structures. It can mean the bony pelvis, a ring of bone between the trunk and the thigh; the lesser or true pelvis, the part of the bony pelvis below the pelvic inlet; or the pelvic cavity, a funnel-shaped region within the lesser pelvis that contains pelvic organs.

Differences occur between the male and female pelvis. The female pelvis is designed for childbirth, and has a shorter canal, and a larger inlet and outlet to allow the passage of the baby's head during childbirth. The male pelvis has a smaller inlet and outlet and a longer canal. The bones are generally larger, with larger joint surfaces, reflecting the generally stronger build and heavier weight of men.

Female Pelvis

With a larger inlet and outlet, and a shorter canal, the female pelvis is designed to support the fetus during pregnancy, and to facilitate childbirth.

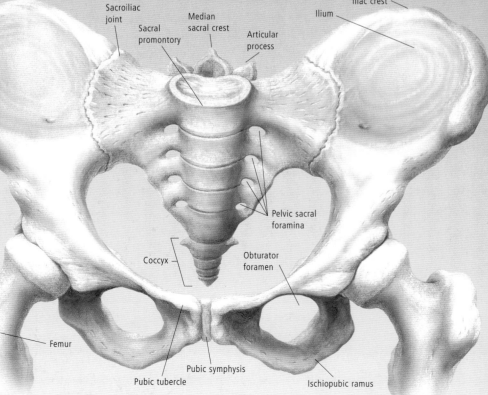

Sacroiliac joint

Sacral promontory

Median sacral crest

Articular process

Iliac crest

Ilium

Anterior superior iliac spine

Pelvic sacral foramina

Coccyx

Obturator foramen

Femur

Pubic symphysis

Pubic tubercle

Ischiopubic ramus

The bony pelvis comprises the hip bones (the ilium, ischium and pubis), sacrum and coccyx. The hip bones are joined at the front by the pubic symphysis. The sacroiliac joints join the hip bone to the sacrum. The muscle-covered bony pelvis transfers weight from the spine to the lower limbs, and provides a protective structure for the abdominal organs.

Open to the abdominal cavity above, the lesser pelvis is bounded behind and above by the sacrum and coccyx. Muscles cover the walls of the lesser pelvis, while the floor is formed by the pelvic diaphragm (pelvic floor).

The pelvic cavity is the area bounded by the pelvic inlet above, and the pelvic diaphragm (pelvic floor) below. Providing protection to the organs within it, the pelvic cavity contains the bladder and rectum; the uterus and vagina in the female; and the prostate and seminal vesicles in the male. The small and large intestines terminate in the pelvic cavity.

PELVIC FLOOR MUSCLES

Muscles line the pelvis, and stretch from the sacrum at the back to the hip bones in the front and sides forming a floor around the pelvic organs. The pelvic floor muscles provide support for the pelvic organs (bladder, rectum, uterus and vagina in women, prostate and seminal vesicles in men) and control the sphincter muscles of the rectum and vagina.

SACROILIAC JOINT

Joining the ilium of the hip bone to the sacrum of the spine is the sacroiliac joint. Strong ligaments link the two bones, to create a stable joint with very limited movement, yet with the ability to cope with various body movements. Immense stresses are placed on this joint, contending not only with the downward pressure of the body's weight, but also with the upward thrust of the legs and pelvis.

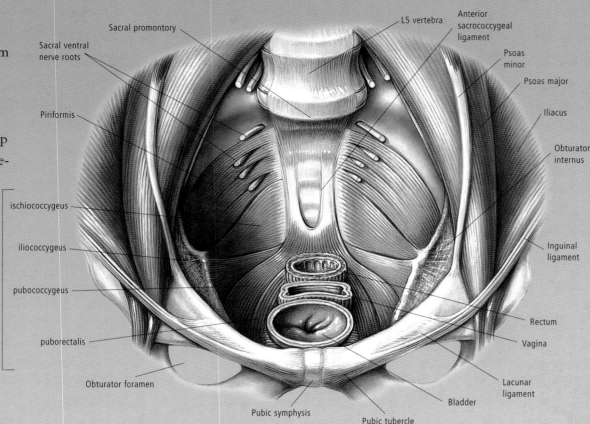

Sacral promontory
Sacral ventral nerve roots
Piriformis
Levator ani
  ischiococcygeus
  iliococcygeus
  pubococcygeus
  puborectalis
Obturator foramen
Pubic symphysis
Pubic tubercle
L5 vertebra
Anterior sacrococcygeal ligament
Psoas minor
Psoas major
Iliacus
Obturator internus
Inguinal ligament
Rectum
Vagina
Lacunar ligament
Bladder

Pelvic floor muscles

The pelvic floor muscles, formed by the coccygeus and the levator ani muscles, stretch from the sacrum at the back to the hip bones at the front, creating a muscular cradle for the pelvic organs.

Sacroiliac joint

The sacroiliac joint articulates between the sacrum and ilium bone of the hip. Great stresses are placed on this joint; it must withstand the weight of the upper body, and cope with body movements. To stabilize the joint, powerful ligaments hold it firm, allowing only limited movement.

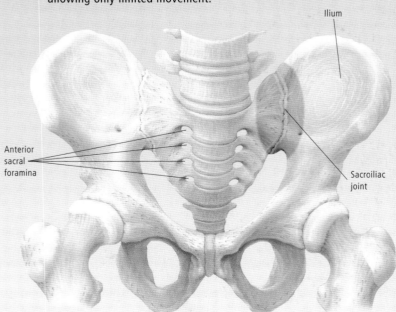

Ilium
Anterior sacral foramina
Sacroiliac joint

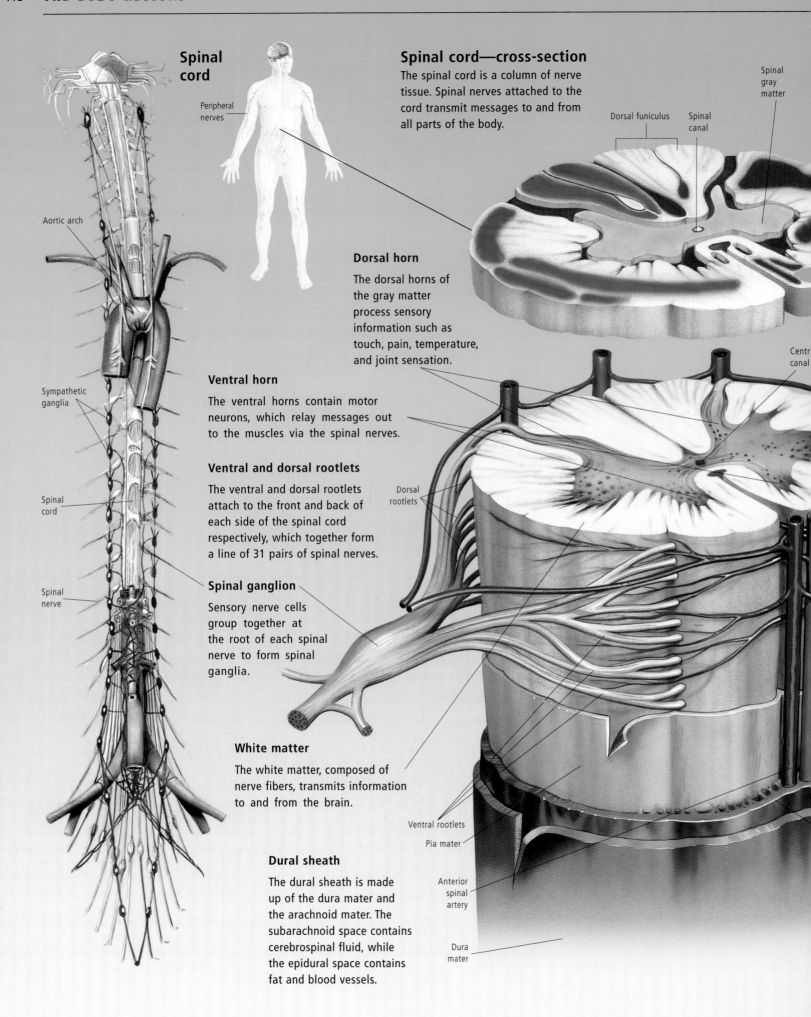

Spinal cord

Peripheral nerves

Aortic arch

Sympathetic ganglia

Spinal cord

Spinal nerve

Spinal cord—cross-section

The spinal cord is a column of nerve tissue. Spinal nerves attached to the cord transmit messages to and from all parts of the body.

Dorsal funiculus

Spinal canal

Spinal gray matter

Centr canal

Dorsal horn

The dorsal horns of the gray matter process sensory information such as touch, pain, temperature, and joint sensation.

Ventral horn

The ventral horns contain motor neurons, which relay messages out to the muscles via the spinal nerves.

Ventral and dorsal rootlets

The ventral and dorsal rootlets attach to the front and back of each side of the spinal cord respectively, which together form a line of 31 pairs of spinal nerves.

Dorsal rootlets

Spinal ganglion

Sensory nerve cells group together at the root of each spinal nerve to form spinal ganglia.

White matter

The white matter, composed of nerve fibers, transmits information to and from the brain.

Ventral rootlets

Pia mater

Dural sheath

The dural sheath is made up of the dura mater and the arachnoid mater. The subarachnoid space contains cerebrospinal fluid, while the epidural space contains fat and blood vessels.

Anterior spinal artery

Dura mater

The Spinal Cord

The spinal cord is a major component of the nervous system, responsible for the relaying of messages to and from the brain. Measuring around 18 inches (45 centimeters) long, the spinal cord runs from the brain down to the lower back, terminating around L1 or L2, and below this it fans out in a network of lumbar and sacral nerve rootlets, known as the cauda equina. Around the spinal cord is a protective sheath (dural sheath) and a fluid layer (cerebrospinal fluid) to provide protection and act as shock absorbers. The spinal cord and its outer layers run through a hollow canal formed by the vertebrae (the spinal or vertebral canal), which surrounds the cord in a strong protective casing.

A central H-shaped core of gray matter, surrounded by white matter, runs through the spinal cord. The nerve cells of the gray and white matter process and transmit sensory information to the brain and motor information from the brain.

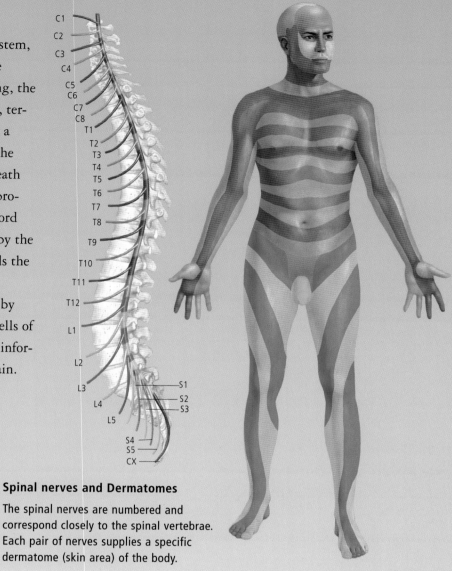

Spinal nerves and Dermatomes

The spinal nerves are numbered and correspond closely to the spinal vertebrae. Each pair of nerves supplies a specific dermatome (skin area) of the body.

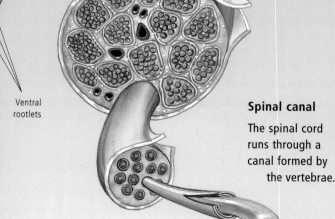

Ventral rootlets

Arachnoid mater

Axon

Myelin sheath of Schwann cell

Spinal canal

The spinal cord runs through a canal formed by the vertebrae.

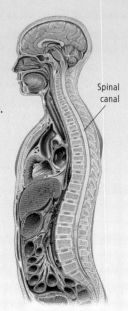

Spinal canal

THE SPINAL CANAL

Surrounding the spinal cord is the vertebral canal (spinal canal). This conduit for the spinal cord and spinal nerve roots is formed by the vertebrae. An opening (foramen) created by the body and vertebral arch of each vertebrae forms a continuous canal to accommodate the spinal cord.

DERMATOMES

Dermatomes are the various regions of the skin controlled by each of the spinal nerves. The 31 pairs of nerves transmit messages in and out of the central nervous system to their respective dermatome.

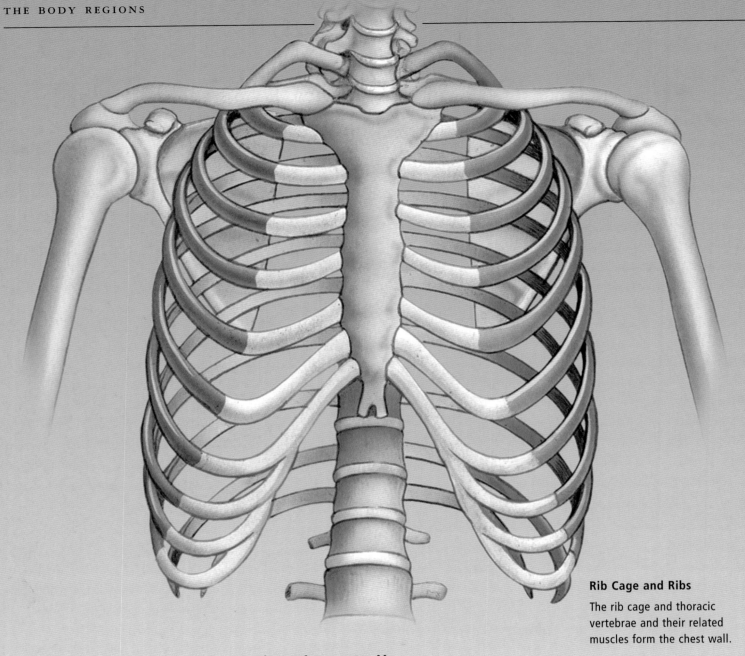

Rib Cage and Ribs

The rib cage and thoracic vertebrae and their related muscles form the chest wall.

The Chest Wall

The chest wall is made up of the rib cage and sternum, thoracic vertebrae and associated musculature, and provides protection for the internal organs of the chest region. The muscles of the chest wall move the rib cage and play a vital role in breathing.

THE RIB CAGE

The rib cage provides a suit of armor for the vital organs contained in the chest cavity. There are 12 pairs of ribs, each joined to a corresponding thoracic vertebrae (T1–12) of the spine. Spanning round to the front, the first seven pairs of ribs, called true ribs, join to the breast bone (sternum). The next three pairs of ribs (pairs 8–10), called false ribs, are not attached to the breast bone, but join each other and then attach to the seventh rib. The remaining two pairs of ribs are not attached at the front and are called floating ribs.

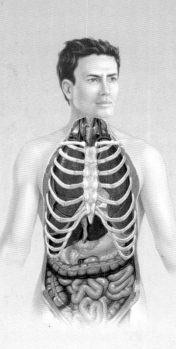

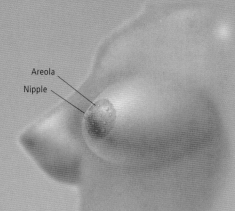

Areola
Nipple

THE NIPPLE

In the center of each breast is the nipple, a raised area encircled by a reddish brown pigmented area called the areola. During breastfeeding, breast milk is channelled from the lobules, along lactiferous ducts, to tiny exits in the nipple.

The nipple also contains some erectile tissue, and can be an erogenous zone, common to both sexes.

Nipple

The nipple is a conical projection of each breast. The areola is the reddish brown region of skin around each nipple.

THE BREAST

A single pair of mammary glands, or breasts, develop under the skin of the upper chest. While throughout life the male mammary glands remain similar to that of prepubescent females, the female breasts undergo change and purpose, during puberty, during pregnancy and after childbirth. Consisting mainly of fat cells and glands called lobules, the breasts are normally only functional in the female. Breast development occurs during puberty, at which time the nipple becomes pigmented, and the amount of fat and fibrous tissue increases.

Triggered by hormones, the mammary glands are activated during pregnancy. In the days immediately following childbirth, colostrum is produced, followed by true milk several days after birth. Milk production is initiated by the pituitary gland, which releases prolactin, a lactation-stimulating hormone. The glandular tissue breaks down after weaning, and the breast returns to its original composition of fat and fibrous tissue.

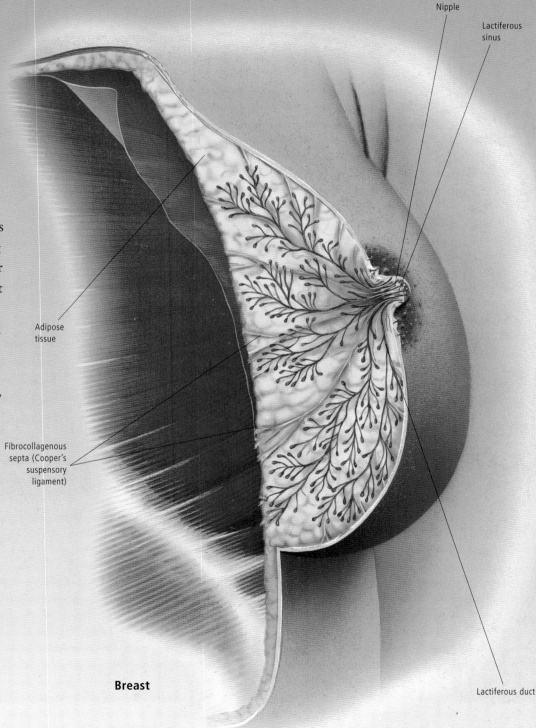

Nipple

Lactiferous sinus

Adipose tissue

Fibrocollagenous septa (Cooper's suspensory ligament)

Lactiferous duct

Breast

The Chest and Chest Cavity

The Chest and Chest Cavity

The internal organs of the chest are protected by a bony suit of armor. The thoracic cage, formed by the ribs, the breastbone (sternum) and the twelve thoracic vertebrae, provides a line of defense against injury and damage to the vital organs of the chest—the heart and lungs.

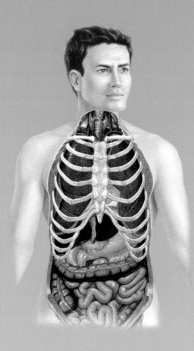

Protection for internal organs

The thoracic cage provides protection to the vital organs and major blood vessels within the chest.

Rib cage

Lungs

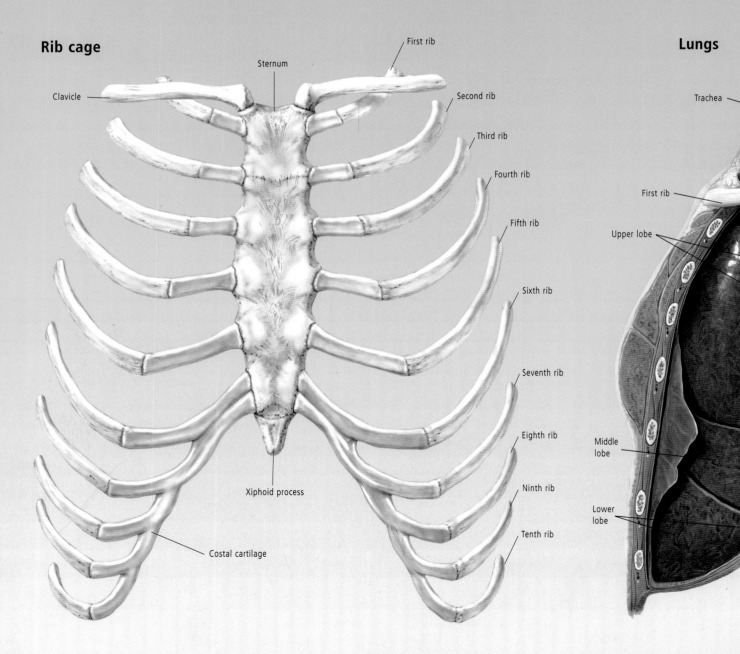

Clavicle

Sternum

First rib

Second rib

Third rib

Fourth rib

Fifth rib

Sixth rib

Seventh rib

Eighth rib

Ninth rib

Tenth rib

Xiphoid process

Costal cartilage

Trachea

First rib

Upper lobe

Middle lobe

Lower lobe

The Heart–Lung Relationship

The heart and lungs work together to maintain our blood supply. As blood circulates through our body, its oxygen is consumed and replaced with carbon dioxide. The right ventricle of the heart sends the deoxygenated blood to the lungs via the pulmonary arteries. These arteries closely follow the bronchial tree and terminate at the capillaries of the tiny air sacs (alveoli), where the carbon dioxide is removed and replaced once more with oxygen. The reoxygenated blood is returned to the left chamber (atrium) of the heart for circulation around the body again.

DID YOU KNOW?

The sound of the heartbeat is created by the opening and closing of the valves. On average, the heart beats 100,000 times per day. In a normal lifetime that is around 2.5 billion heartbeats.

Lungs and heart

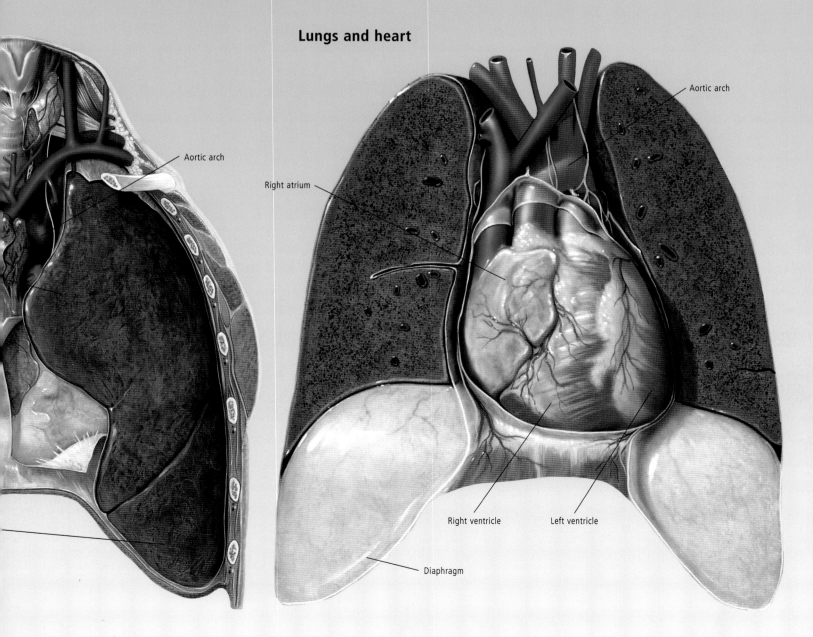

Aortic arch

Aortic arch

Right atrium

Right ventricle

Left ventricle

Diaphragm

Heart—front

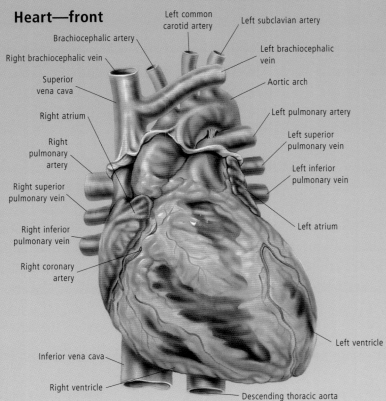

Brachiocephalic artery

Left common carotid artery

Left subclavian artery

Right brachiocephalic vein

Left brachiocephalic vein

Superior vena cava

Aortic arch

Right atrium

Left pulmonary artery

Right pulmonary artery

Left superior pulmonary vein

Left inferior pulmonary vein

Right superior pulmonary vein

Right inferior pulmonary vein

Left atrium

Right coronary artery

Left ventricle

Inferior vena cava

Right ventricle

Descending thoracic aorta

Heart—section showing all four chambers

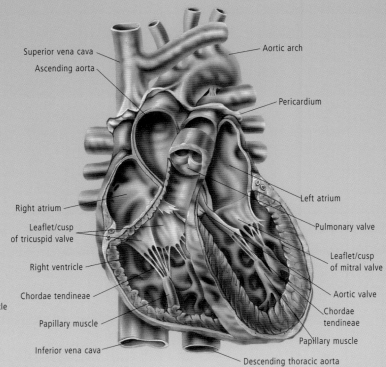

Superior vena cava

Aortic arch

Ascending aorta

Pericardium

Right atrium

Left atrium

Leaflet/cusp of tricuspid valve

Pulmonary valve

Right ventricle

Leaflet/cusp of mitral valve

Chordae tendineae

Aortic valve

Papillary muscle

Chordae tendineae

Inferior vena cava

Papillary muscle

Descending thoracic aorta

The Heart

Cardiac muscle

THE ANATOMY OF THE HEART

The heart is a muscular pump, about the size of a fist, located between the lungs. Divided into two halves by the muscular wall of the septum, each half is divided again into an upper and lower chamber. On the left of the heart are the left atrium and left ventricle responsible for receiving and circulating oxygenated blood from the lungs. On the right of the heart are the right atrium and right ventricle, responsible for receiving deoxygenated blood and circulating it, via the pulmonary arteries, to the lungs, where gas exchange occurs. Each atrium is separated from its related ventricle by a valve which allows blood flow in one direction only; the mitral valve lies between the left atrium and left ventricle, and the tricuspid valve lies between the right atrium and right ventricle. Tough fibrous cords, the chordae tendineae or "heart strings", extend from the lower free edge of the valves and connect to papillary muscles. Two more valves lie at the exits of the ventricles; the pulmonary valve lies at the outlet from the right ventricle, while the body aortic valve lies at the outlet from the left ventricle. These valves perform a similar function to the atrioventricular valves, allowing blood flow out of the ventricles only.

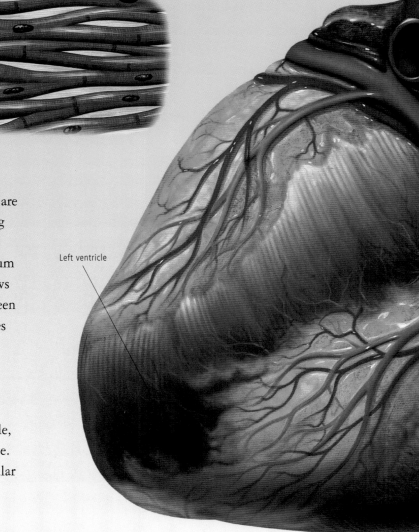

Left ventricle

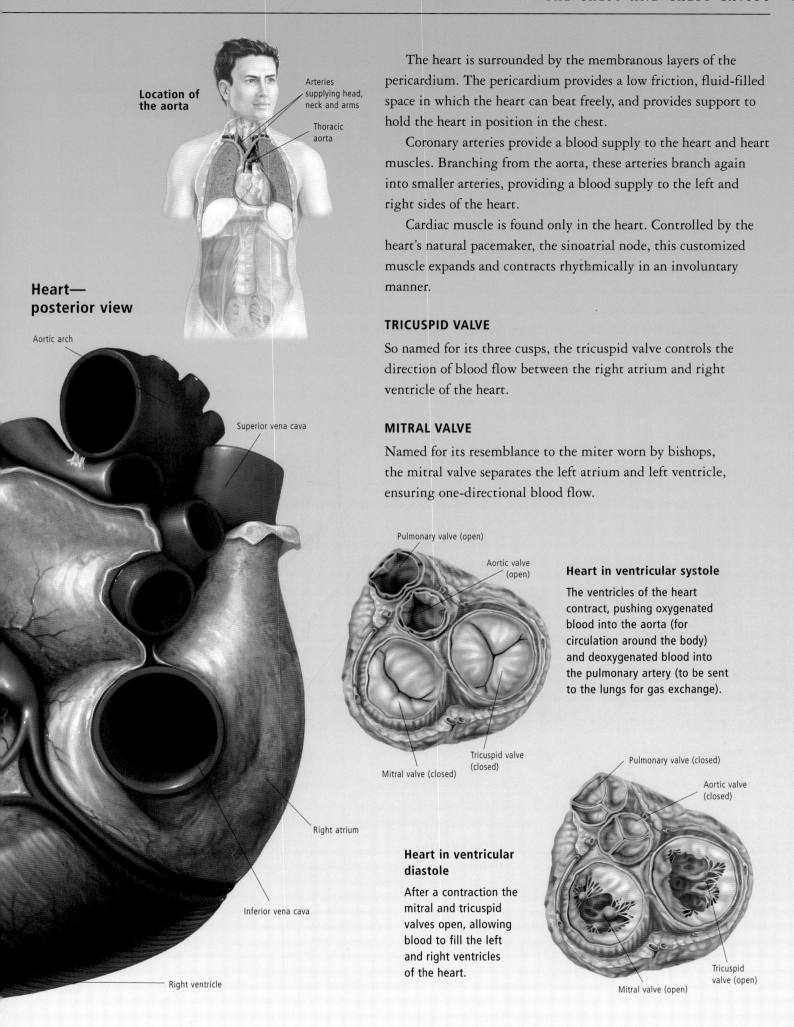

Location of the aorta

Arteries supplying head, neck and arms

Thoracic aorta

Heart— posterior view

Aortic arch

Superior vena cava

Right atrium

Inferior vena cava

Right ventricle

The heart is surrounded by the membranous layers of the pericardium. The pericardium provides a low friction, fluid-filled space in which the heart can beat freely, and provides support to hold the heart in position in the chest.

Coronary arteries provide a blood supply to the heart and heart muscles. Branching from the aorta, these arteries branch again into smaller arteries, providing a blood supply to the left and right sides of the heart.

Cardiac muscle is found only in the heart. Controlled by the heart's natural pacemaker, the sinoatrial node, this customized muscle expands and contracts rhythmically in an involuntary manner.

TRICUSPID VALVE

So named for its three cusps, the tricuspid valve controls the direction of blood flow between the right atrium and right ventricle of the heart.

MITRAL VALVE

Named for its resemblance to the miter worn by bishops, the mitral valve separates the left atrium and left ventricle, ensuring one-directional blood flow.

Pulmonary valve (open)

Aortic valve (open)

Heart in ventricular systole

The ventricles of the heart contract, pushing oxygenated blood into the aorta (for circulation around the body) and deoxygenated blood into the pulmonary artery (to be sent to the lungs for gas exchange).

Mitral valve (closed)

Tricuspid valve (closed)

Pulmonary valve (closed)

Aortic valve (closed)

Heart in ventricular diastole

After a contraction the mitral and tricuspid valves open, allowing blood to fill the left and right ventricles of the heart.

Mitral valve (open)

Tricuspid valve (open)

Heart Cycle

In the cardiac cycle, the chambers of the heart pass through a relaxation phase (diastole) and a contraction phase (systole). At the beginning of each contraction the atrioventricular valves close and the pulmonary and aortic valves open. At the end of each contraction the pulmonary and aortic valves close and the atrioventricular valves open.

HEARTBEAT

Regulated by the sinoatrial node, the contraction of the ventricles produces the heartbeat which can be felt on the lower left side of the chest. The sinoatrial node is known as the heart's natural pacemaker, and is located in the right atrium. This node transmits electrical impulses that determine the rate and rhythm of the contractions of the heart, producing the heartbeat. These heartbeats create a pulse which can be monitored at the pulse points of the body.

Heart cycle

In the cardiac cycle, the chambers of the heart pass through a relaxation phase (diastole) and a contraction phase (systole).

Heart cycle 1

In atrial diastole (at the beginning of ventricular diastole), deoxygenated blood from the systemic circulation and oxygenated blood from the lungs enter the left and right chambers (atria) of the heart respectively.

PULMONARY ARTERY

The pulmonary valve opens to release deoxygenated blood from the right ventricle of the heart into the pulmonary artery. This blood is then transported along the pulmonary arteries to the lungs where carbon dioxide is removed and oxygen is replenished.

SYSTEMIC AND PULMONARY CIRCULATION

Systemic circulation takes the blood around the body and back to the heart. On its journey through the body, the oxygen and nutrients are extracted by the body, and carbon dioxide is collected in the blood. This blood then returns to the heart, entering via the right atrium. After passing through the tricuspid valve, this blood enters the right ventricle, which pumps it through the pulmonary valve and into the pulmonary arteries. These arteries carry the deoxygenated blood to the lungs. The arteries branch out repeatedly, until they become tiny capillaries around the alveoli. The alveoli are tiny air sacs which extract carbon dioxide from the blood; this carbon dioxide is subsequently exhaled by the lungs. The alveoli replace the oxygen, provided by inhaled air, and the freshly oxygenated blood is returned to the heart via the

Sinoatrial node

Heartbeat

The heart's natural pacemaker, the sinoatrial node, sends out electrical impulses that control the rate and rhythm of the heartbeat.

Superior vena cava

Ascending aorta

Right pulmonary artery

Right superior pulmonary vein

Right inferior pulmonary vein

Right atrium

Inferior vena cava

Left pulmonary artery

Left superior pulmonary vein

Left inferior pulmonary vein

Left atrium

Mitral valve

Left ventricle

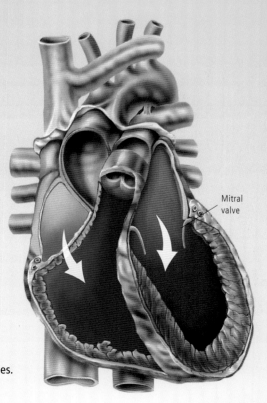

Heart cycle 2

Towards the end of ventricular diastole, the atria contract (atrial systole) and pump blood into the left and right ventricles.

Mitral valve

Pulmonary circulation

After a circuit of the body (systemic circulation), oxygen-depleted blood is pumped from the heart to the lungs (pulmonary circulation), where the oxygen is replenished and then pumped back to the heart in a continuous cycle.

Pulmonary artery

Deoxygenated blood from the right ventricle of the heart is carried through the pulmonary trunk, or pulmonary artery, towards the lungs.

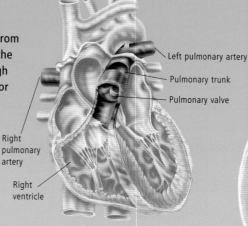

Left pulmonary artery

Pulmonary trunk

Pulmonary valve

Right pulmonary artery

Right ventricle

capillaries, then venules, then into pulmonary veins, which pump the blood into the left atrium. From here the blood passes through the mitral valve into the left ventricle and then into the body via the systemic circulation, repeating the cycle once again. The complete cycle, pumping about 10 pints (almost 5 liters) of blood around the body at rest, takes about one minute.

Oxygenated blood flows out of the lungs to the left side of the heart and is pumped out into the body for systemic circulation.

Oxygen-depleted blood enters the right ventricle of the heart and is pumped into the lungs to be oxygenated by the alveoli.

NB: In this illustration the top two-thirds of the lungs and pleura have been cut away to show the heart.

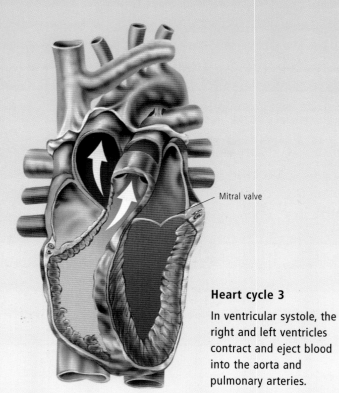

Mitral valve

Heart cycle 3

In ventricular systole, the right and left ventricles contract and eject blood into the aorta and pulmonary arteries.

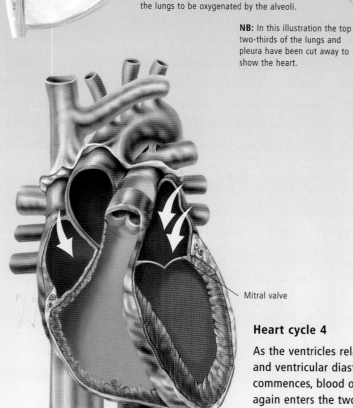

Mitral valve

Heart cycle 4

As the ventricles relax and ventricular diastole commences, blood once again enters the two atria and the cycle begins again.

The Lungs

The lungs are paired organs that are responsible for gas exchange between the atmosphere and the blood. Inhaled oxygen is supplied to the blood, and carbon dioxide removed from the blood is exhaled. Each of the lungs is divided into lobes: the left lung has two lobes, and the right lung has three lobes. The lobed lungs are enclosed in a double-layered membrane called the pleura. This membrane allows the lungs to move against the ribcage with minimal friction. Lying between the two lungs is the mediastinum, containing the heart, esophagus, trachea and major vessels and nerves. The positioning of these organs results in the left lung being smaller than the right, as the heart and its associated vessels require more space in the left side of the chest.

The trachea branches into the two major airways of the left and right main bronchi; these two airways then branch into smaller bronchi, which again branch into even smaller bronchioles. The bronchioles then subdivide into tiny clusters of air sacs (alveoli). This network of bronchi and bronchioles is known as the bronchial tree.

Lungs

The two lungs are divided into lobes by a series of clefts or fissures. Inhaled air is carried down the trachea into the bronchial tree within the lungs, where oxygen and carbon dioxide are exchanged.

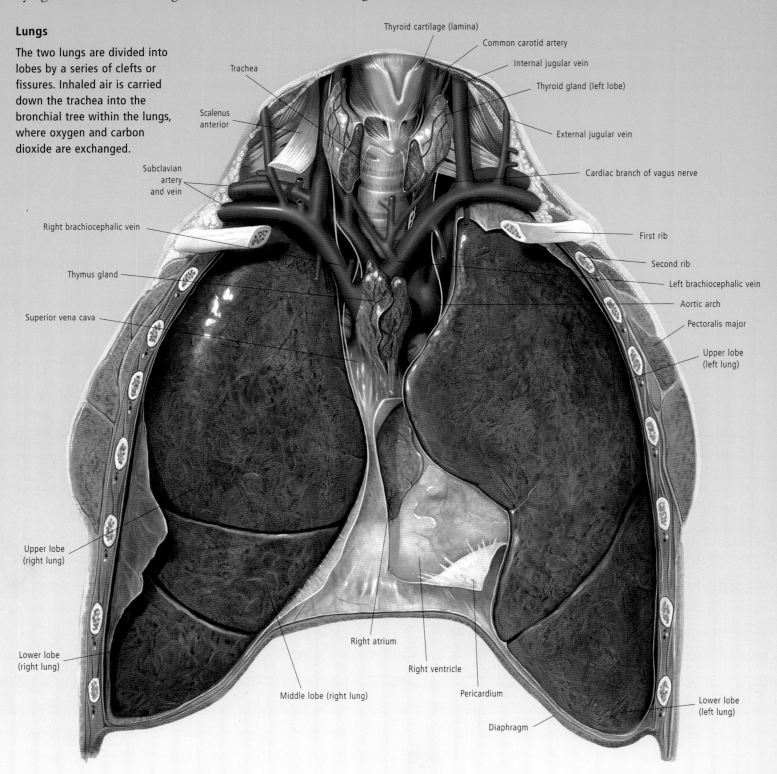

Thyroid cartilage (lamina)

Common carotid artery

Internal jugular vein

Thyroid gland (left lobe)

External jugular vein

Trachea

Scalenus anterior

Cardiac branch of vagus nerve

Subclavian artery and vein

First rib

Right brachiocephalic vein

Second rib

Left brachiocephalic vein

Thymus gland

Aortic arch

Superior vena cava

Pectoralis major

Upper lobe (left lung)

Upper lobe (right lung)

Lower lobe (right lung)

Right atrium

Right ventricle

Lower lobe (left lung)

Middle lobe (right lung)

Pericardium

Diaphragm

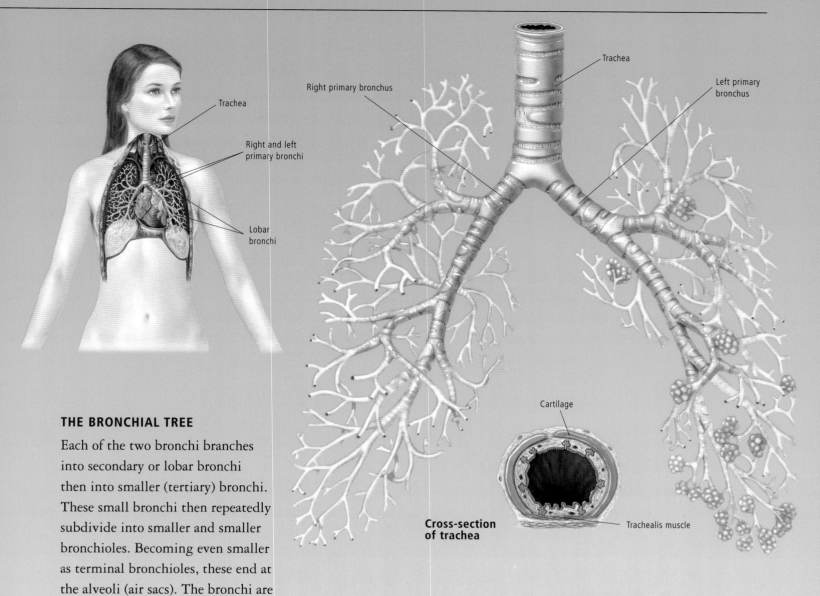

Trachea

Right primary bronchus

Left primary bronchus

Trachea

Right and left primary bronchi

Lobar bronchi

Cartilage

Cross-section of trachea

Trachealis muscle

THE BRONCHIAL TREE

Each of the two bronchi branches into secondary or lobar bronchi then into smaller (tertiary) bronchi. These small bronchi then repeatedly subdivide into smaller and smaller bronchioles. Becoming even smaller as terminal bronchioles, these end at the alveoli (air sacs). The bronchi are strengthened by cartilage, while the bronchioles are entirely muscular.

Bronchial tree

The bronchial tree is a network of airways carrying inhaled air to the tiny air sacs (alveoli) where gas exchange occurs. To ensure the airways remain clear, the bronchi are lined with mucus membrane and cilia, minute hair-like structures, that trap dust and particles.

Cilia

Cross-section of bronchi

Bronchial gland

Breathing

Breathing (respiration or ventilation) is the inspiration and expiration of air into and out of the lungs, and the process of gas exchange.

Breathing muscles

The movement of the intercostal muscles, in conjunction with the diaphragm, increases the capacity of the rib cage during inspiration.

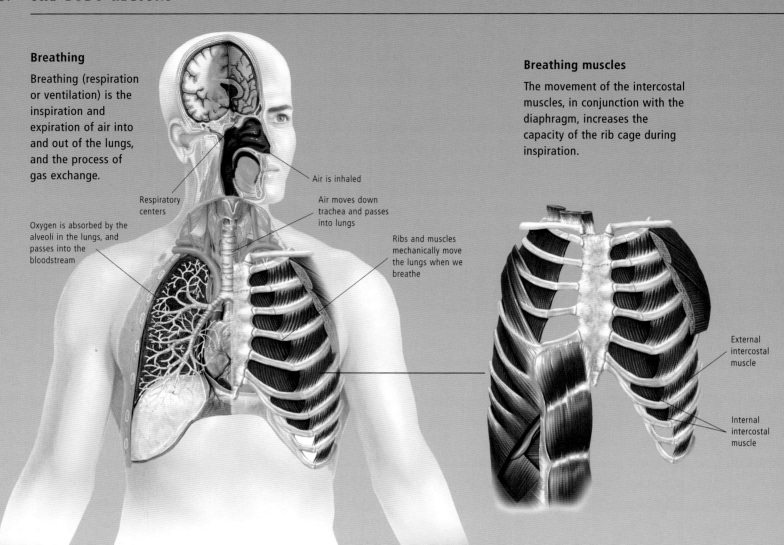

Respiratory centers

Air is inhaled

Air moves down trachea and passes into lungs

Oxygen is absorbed by the alveoli in the lungs, and passes into the bloodstream

Ribs and muscles mechanically move the lungs when we breathe

External intercostal muscle

Internal intercostal muscle

Breathing

HOW WE BREATHE

Taking a breath of air (known as inspiration) triggers action in many parts of the chest. The rib cage, raised by its layers of intercostal muscle, expands slightly, and below the lungs the diaphragm moves down. This movement increases the capacity of the thoracic cage, allowing room for the lungs to expand when air is inhaled. Air, inhaled through the mouth or nose, passes through the pharynx, larynx, trachea, bronchial airways, and finally reaches the alveoli, where the oxygen is exchanged for carbon dioxide, a waste product of the blood. The carbon dioxide is then exhaled (known as expiration). In expiration the carbon dioxide is forced out of the lungs as the ribs recoil and the diaphragm moves up, reducing the capacity of the thoracic cage. As the lungs compress and expand, the pressure within also rises and falls in relation to outside atmospheric pressure.

Alveoli

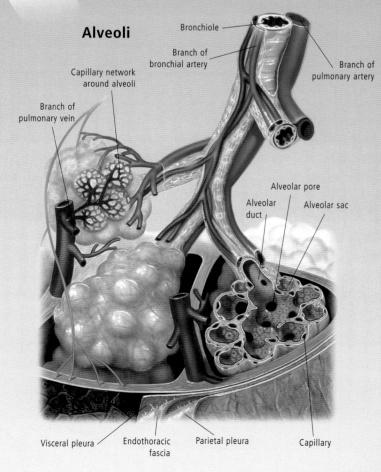

Bronchiole

Branch of bronchial artery

Capillary network around alveoli

Branch of pulmonary artery

Branch of pulmonary vein

Alveolar pore

Alveolar duct

Alveolar sac

Visceral pleura

Endothoracic fascia

Parietal pleura

Capillary

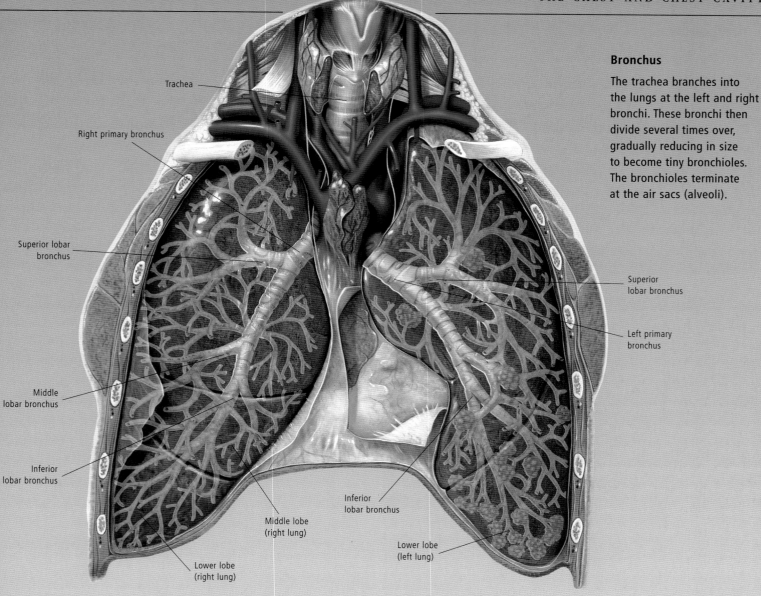

Trachea

Right primary bronchus

Superior lobar
bronchus

Middle
lobar bronchus

Inferior
lobar bronchus

Lower lobe
(right lung)

Middle lobe
(right lung)

Inferior
lobar bronchus

Lower lobe
(left lung)

Superior
lobar bronchus

Left primary
bronchus

Bronchus

The trachea branches into
the lungs at the left and right
bronchi. These bronchi then
divide several times over,
gradually reducing in size
to become tiny bronchioles.
The bronchioles terminate
at the air sacs (alveoli).

Breathing

When we breathe in, the intercostal muscles contract and the
diaphragm moves down. This movement increases the capacity
of the chest cavity to accommodate the intake of air into the
expanding lungs. After gas exchange occurs, the breathing muscles
recoil (a passive process), returning to their original positions.

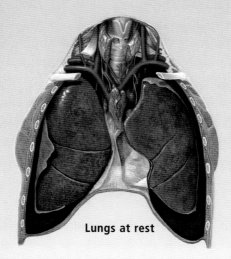

Lungs at rest

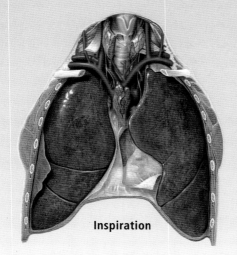

Inspiration

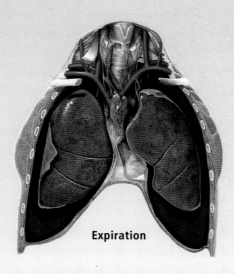

Expiration

The Chest Cavity

The ribs and spine join at the back of the thoracic cage. Between the ribs are layers of intercostal muscles that move the rib cage during breathing. The smooth, frictionless movement between the ribs and the lungs is made possible by the double-layered membrane wrapping around the lungs, called the pleura. The outer layer of the pleura, the parietal layer, is attached to the rib cage.

The Diaphragm

The diaphragm is the main muscle used to draw air into the body and it is controlled by the phrenic nerve. At rest, the diaphragm is dome-shaped, but when air is taken in by the body, the diaphragm contracts, moving downwards, thereby increasing the capacity of the chest cavity. It returns to its original position when the chest and abdomen relax. A muscular layer, the diaphragm is attached to the thoracic cage, joining with the spine at the back, the ribs at the side of the chest, and the breastbone (sternum) at the front of the chest. The central tendon of the diaphragm is attached to the pericardial sac, which surrounds the heart. The diaphragm separates the chest and abdominal cavities, and several structures pass through it, including the esophagus and major blood vessels of the circulatory system.

The Esophagus

A component of the digestive system, the esophagus runs through the neck and the chest, passing through the diaphragm on its way to the stomach. Food and liquids travel down this passageway. At the top, where the esophagus joins with the pharynx, and at the bottom, where it joins with the stomach, are strong sphincter muscles. These muscles expand and contract to open and close the entrance and exit to the esophagus. The esophagus is a tube of smooth muscle which moves in wave-like contractions to move food down to the stomach. The muscles of the esophagus are involuntary muscles controlled by the autonomic nervous system.

The Trachea

Running from the neck into the chest cavity, the trachea sits immediately in front of the upper section of the esophagus, before branching into the two main bronchi. The trachea is a muscular structure strengthened by C-shaped cartilages.

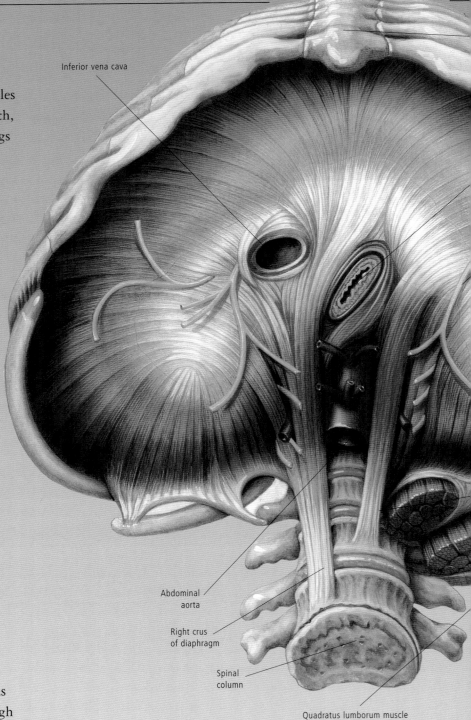

Inferior vena cava

Abdominal aorta

Right crus of diaphragm

Spinal column

Quadratus lumborum muscle

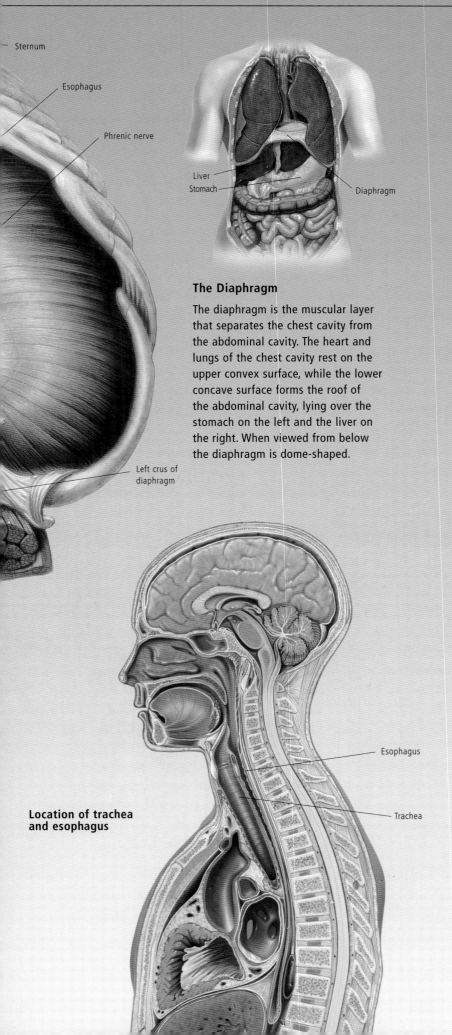

Sternum

Esophagus

Phrenic nerve

Liver

Stomach

Diaphragm

Left crus of
diaphragm

The Diaphragm

The diaphragm is the muscular layer that separates the chest cavity from the abdominal cavity. The heart and lungs of the chest cavity rest on the upper convex surface, while the lower concave surface forms the roof of the abdominal cavity, lying over the stomach on the left and the liver on the right. When viewed from below the diaphragm is dome-shaped.

Esophagus

Trachea

**Location of trachea
and esophagus**

The Thymus Gland

Found behind the breastbone (sternum), the thymus is not really a gland, but a lymph organ responsible for the production of special lymphocytes used to defend the body against foreign cells such as viruses. Active during childhood and adolescence, the thymus weighs about $^1/_2$ ounce (14 grams) at birth, increasing to about 1 ounce (28 grams) at puberty. By adulthood the thymus has reduced back to its original weight. During its active period, it produces lymphocytes called thymus-derived or "T" cells. T cells take about 3 weeks to develop and are then released into the bloodstream. Circulated to lymph tissue in the body, the function of the T cells is to recognize and destroy foreign cells. Once a foreign cell has been identified, the T cells multiply, attack and destroy the invader.

Thymus

The thymus is a ductless gland, found in the chest behind the sternum. The thymus produces a special type of white blood cell, the T lymphocyte, an important element in cell-mediated immune response.

Thymus

The Abdomen

The Abdomen

The lower and larger of the two body cavities (the other being the chest or thoracic cavity), the abdominal cavity houses some of the major organs of the body. These include the main organs of the digestive system, the urinary system and the reproductive system. Digestive system organs within the abdomen include the stomach, liver, gallbladder, pancreas and intestine. Major organs of the urinary system and reproductive system are also found in the abdominal cavity but have been allocated their own chapter in this book.

Also running through the abdominal cavity are major blood vessels leading to and from the heart; the descending aorta leads to the iliac arteries that supply the pelvis and lower limbs; blood from the lower limbs returns to the heart via the inferior vena cava, one of the largest veins in the body.

Surrounding the organs and lining the walls of the abdomen is a lubricating layer of membrane, the peritoneum. This membrane layer allows the muscular movements of the intestines during the digestive process.

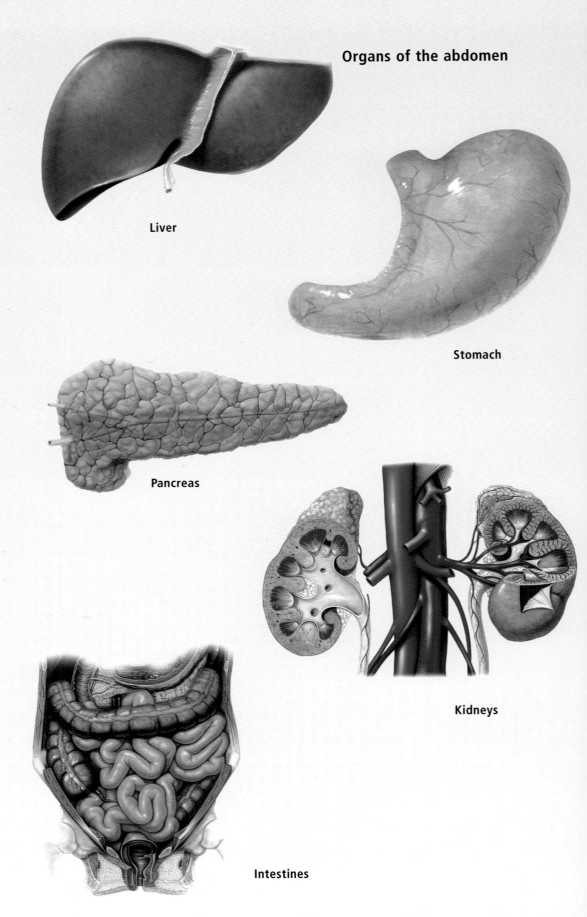

Organs of the abdomen

Liver

Stomach

Pancreas

Kidneys

Intestines

Organs of the abdomen

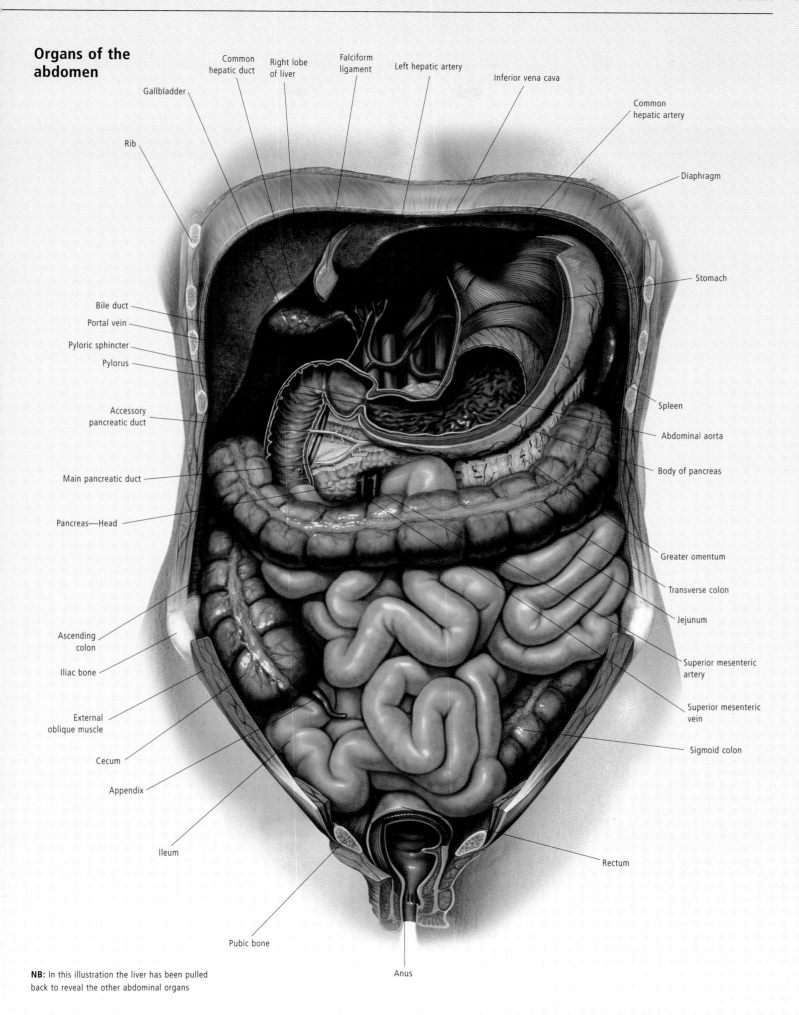

Common hepatic duct

Right lobe of liver

Falciform ligament

Left hepatic artery

Inferior vena cava

Common hepatic artery

Gallbladder

Rib

Diaphragm

Bile duct

Portal vein

Pyloric sphincter

Pylorus

Stomach

Spleen

Accessory pancreatic duct

Abdominal aorta

Body of pancreas

Main pancreatic duct

Pancreas—Head

Greater omentum

Transverse colon

Jejunum

Ascending colon

Iliac bone

Superior mesenteric artery

External oblique muscle

Superior mesenteric vein

Cecum

Sigmoid colon

Appendix

Ileum

Rectum

Pubic bone

Anus

NB: In this illustration the liver has been pulled back to reveal the other abdominal organs

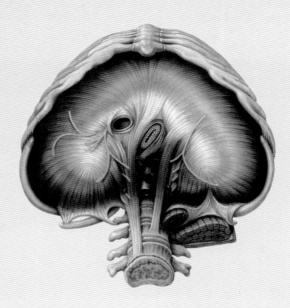

Diaphragm

Separating the chest cavity and the abdominal cavity is the diaphragm, a muscular layer crucial to the breathing process. At rest, it forms a dome-shaped roof of the abdominal cavity. During inspiration (breathing in), the diaphragm contracts, pushing down into the abdomen, thus creating increased capacity in the chest cavity for the intake of air.

The Abdominal Cavity

The abdominal cavity holds its major organ components in a compact yet flexible housing. The back wall of the cavity is formed by the bones of the spine and their associated muscles. The side and front walls are comprised of strong muscle layers covered by fat and skin; these muscles compress the abdominal organs to allow for expansion of the lungs during breathing. The diaphragm forms the roof of the abdominal cavity, while the floor is formed by the pelvic floor (pelvic diaphragm).

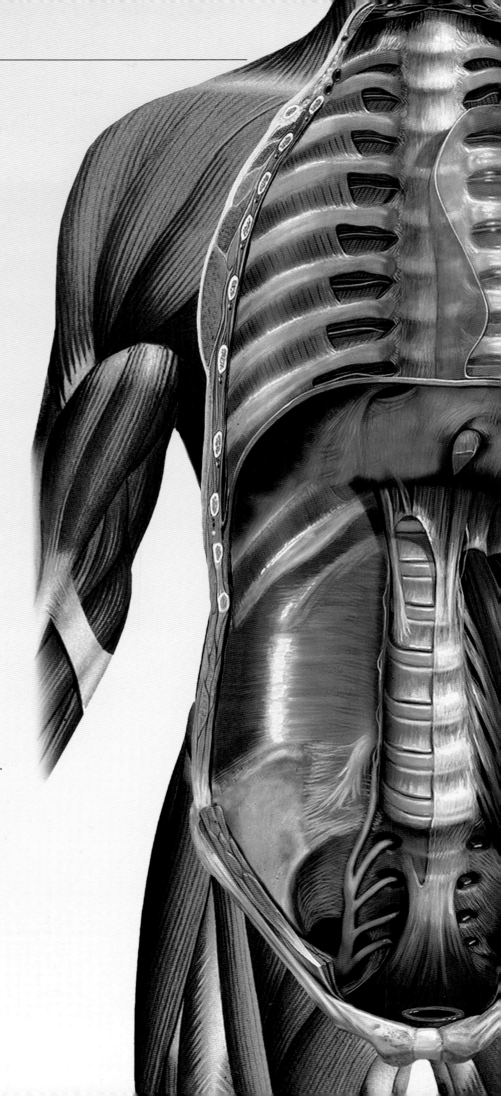

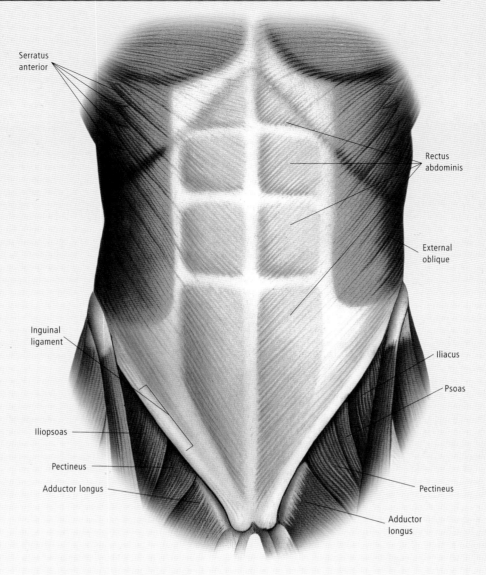

Serratus anterior

Rectus abdominis

External oblique

Iliacus

Psoas

Inguinal ligament

Iliopsoas

Pectineus

Pectineus

Adductor longus

Adductor longus

Muscles of the abdomen

The front section of the abdomen is strengthened by muscular layers, providing a muscular casing for the abdominal organs and assisting in compression of these organs during the breathing process. The external oblique muscle forms the outer layer, the internal oblique muscle forms the middle layer and the transverse abdominis muscle forms the inner layer.

Pelvic floor muscles

The pelvic floor (pelvic diaphragm) forms the floor of the abdominal cavity. This layer of muscles spans across the pelvic area, supporting the abdominal organs. The main muscles of the pelvic floor are the coccygeus and the levator ani muscles.

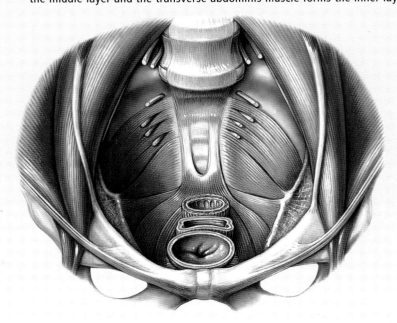

Alimentary canal

The breakdown of food into fats, carbohydrates and proteins takes place in the alimentary canal (digestive tract). The nutrients are absorbed into the blood to provide energy, and to build and maintain body tissues; water is absorbed along the final part of the canal, and the remaining waste is excreted as feces.

Esophagus

The esophagus is a tube of smooth muscle extending from the mouth to the stomach. Muscular contractions of the esophagus create wave-like movements which propel chewed food down the esophagus. Sphincter muscles at the end of the esophagus release the food into the stomach.

Stomach

Hydrochloric acid and enzymes in the stomach act to break down food into smaller components. Comprised of several muscular layers, the muscles churn the food, breaking it down into a mix of food, acid and gastric juices, known as chyme. This chyme is stored, then gradually released into the duodenum via the pylorus and pyloric sphincter.

Duodenum

In the duodenum, bile from the liver and digestive enzymes from the pancreas are added to the chyme. These additional juices further break down the chyme. Extraction of nutrients begins in the duodenum, before the digested food moves on to the remainder of small intestine.

Small intestine

The nutrients from the processed chyme are extracted through the lining of the small intestine and absorbed into the bloodstream for use by the body.

Colon

The remaining mixture entering the large intestine is waste material. Water in the mixture is extracted, and all that remains is feces.

Rectum

Feces are stored in the rectum, to be periodically expelled through the anus.

The Alimentary Canal

Running its course through the upper body, the alimentary canal, or digestive tract, begins at the mouth and ends at the anus. Measuring up to 30 feet (9 meters) in length, the tract is comprised of smooth muscle, which produces a wave-like movement to propel food along the digestive tract, in a process known as peristalsis. In this process, food entering the mouth travels through the esophagus, stomach, duodenum, small intestine, large intestine and rectum. On the journey, it is processed, digested, and broken down into smaller and smaller components. Then the nutrients are extracted in the intestinal tract and absorbed into the bloodstream, to provide the body with energy and to aid tissue repair and body maintenance. Any remaining water is then extracted, until all that remains is waste; this waste is then expelled through the anus.

Stomach function

Arrival in stomach

Even before food arrives in the stomach, the production of gastric juices has commenced, triggered by our various senses of sight, taste, smell, and sometimes, even the thought of food. Food enters the stomach via the junction with the esophagus, known as the cardia, and once in the stomach, the food mixes with the gastric juices.

Gastroesophageal junction

Food matter

Gastric juices mix with food

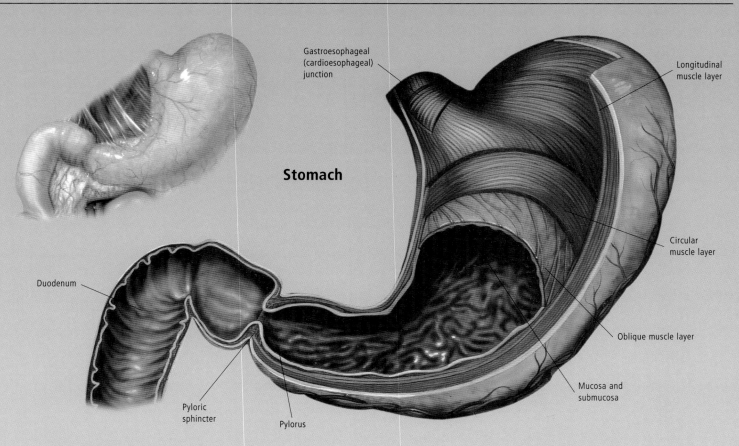

Gastroesophageal
(cardioesophageal)
junction

Longitudinal
muscle layer

Stomach

Circular
muscle layer

Duodenum

Oblique muscle layer

Mucosa and
submucosa

Pyloric
sphincter

Pylorus

The Stomach

THE STRUCTURE AND FUNCTION OF THE STOMACH

Connecting at the top to the esophagus by the cardia (cardioesophageal junction) and at the bottom to the duodenum by the pyloric sphincter, the stomach is made up of several muscular layers. Hormones stimulate the production of stomach acids, including hydrochloric acid, and enzymes, including pepsin. The innermost layer of the stomach, the mucosa and submucosa, produces mucus which provides the

stomach with a protective barrier against its own acids. The muscular layers of the stomach contract in wave-like movements, thereby mixing the food and gastric juices, and breaking it down into a semi-liquid consistency, known as chyme. Once processing in the stomach is completed, muscular contractions push the chyme into the pyloric canal at the end of the stomach. The sphincter muscle at the end of the pyloric canal releases the partially digested mixture into the duodenum.

Digestion

As the acids and enzymes in the stomach break down the food, muscular contractions of the wall of the stomach mix the contents, converting the food and gastric juices into a semi-liquid substance known as chyme.

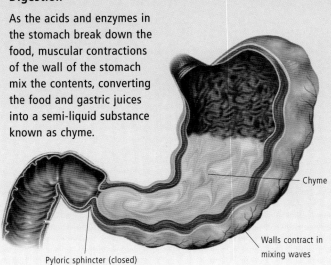

Chyme

Pyloric sphincter (closed)

Walls contract in
mixing waves

Exiting the stomach

Processing of the stomach contents can take several hours. As the process nears completion, the contractions of the stomach decrease, prompting the pyloric sphincter to open, releasing small amounts of chyme into the duodenum.

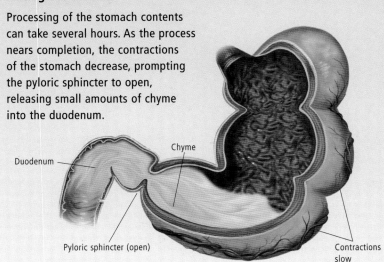

Chyme

Duodenum

Pyloric sphincter (open)

Contractions
slow

Duodenum

The duodenum is the first section of the small intestine, joining the stomach to the jejunum. It lies curved around the head of the pancreas, and measures about 10 inches (25 centimeters) long. The duodenum receives bile from the liver and enzymes from the pancreas; these combine to further process and break down the chyme received from the stomach. Its folded internal surface greatly increases its surface area, allowing the absorption of nutrients such as sugars, fats and amino acids to take place. Digested food is pushed along to the jejunum by contractions of the smooth muscle.

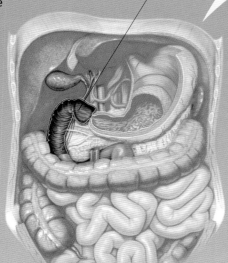

Duodenum

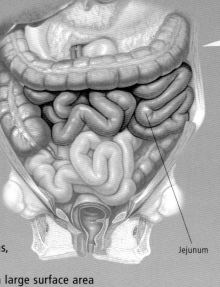

Jejunum

Jejunum

The jejunum is the middle section of the small intestine, and measures about 8 feet (2.5 meters) long. The folds of the internal layer are covered in villi, tiny finger-like projections, themselves covered in microvilli, which create a large surface area for the absorption of nutrients. Nutrients pass through the lining of the walls of the jejunum into the lymphatic vessels and hepatic portal vein to the liver.

The Intestines

The intestines consist of two parts: the small intestine (duodenum, jejunum and ileum) and the large intestine (colon, rectum and anus). The large intestine runs around the margins of the lower half of the abdominal cavity, with the small intestines coiled within.

Lying in the lower section of the alimentary canal, the final stages of the digestive process take place in the intestines. Digested food from the stomach, known as chyme, enters the small intestines, where nutrients from the mix are extracted through the lining of the intestines and are then absorbed into the bloodstream, leaving only waste material to pass through the large intestine. Water and electrolytes are absorbed by the large intestine, leaving only feces, which are stored in the rectum and periodically expelled by the anus.

The intestines are held in place by a layer of membrane called the mesentery. Blood vessels, nerves and lymphatic vessels travel through the mesentery to supply the small intestine.

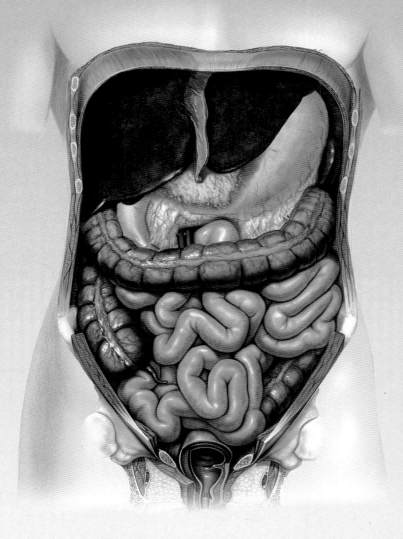

Ileum

Measuring about 12 feet (3.5 meters) long, the ileum is the last section of the small intestine. The ileum continues on from the jejunum and joins to the cecum, part of the large intestine. While most of the nutrients required for body maintenance have been absorbed through the duodenum and jejunum, the ileum serves to absorb bile acids, which it returns to the liver.

Ileum

Colon and appendix

The colon is the first part of the large intestine. Measuring about $4\frac{1}{2}$ feet (1.3 meters) long, it is composed of the cecum, ascending colon, transverse colon, descending colon and sigmoid colon. Joined to the small intestine by the ileum, the colon serves to remove water and salts from the waste material passed on from the small intestine. The appendix is a thin, worm-shaped pouch, $3\frac{1}{2}$ inches (9 centimeters) long. Though attached to the colon, the appendix serves no function.

Transverse colon

Ascending colon

Cecum

Appendix

Sigmoid colon

Rectum

Joining the colon is the rectum, the second last part of the large intestine. Measuring 6–8 inches (15–20 centimeters) long, the rectum receives fecal material from the sigmoid colon and stores it for a short time until it is convenient to expel the stool.

Rectum

Anus

The final part of the large intestine, the anus is a short tube $1\frac{1}{2}$ inches (3–4 centimeters) long that leads from the rectum through the anal sphincter to the anal orifice, through which feces are evacuated.

DID YOU KNOW?

During the course of a year, we each consume around 1100 pounds (500 kilograms) of food. On a daily basis, our digestive system processes around 2–$2\frac{1}{2}$ gallons (10–12 liters) of food, liquids and digestive juices.

Anus

Inside the Intestines

While the intestines measure up to 25 feet (7.5 meters) long, their internal surface area is greatly increased by the folded lining (plicae circularis). Covering this folded lining are more folds called villi. These folds within folds combine to create a huge surface area for the absorption of essential nutrients from digested food. The absorbed nutrients pass into the blood vessels and lymph vessels to be used for cell and tissue maintenance and to provide energy to the body.

Intestines

The final section of the digestive system, the small and large intestines lie in the lower part of the abdominal cavity.

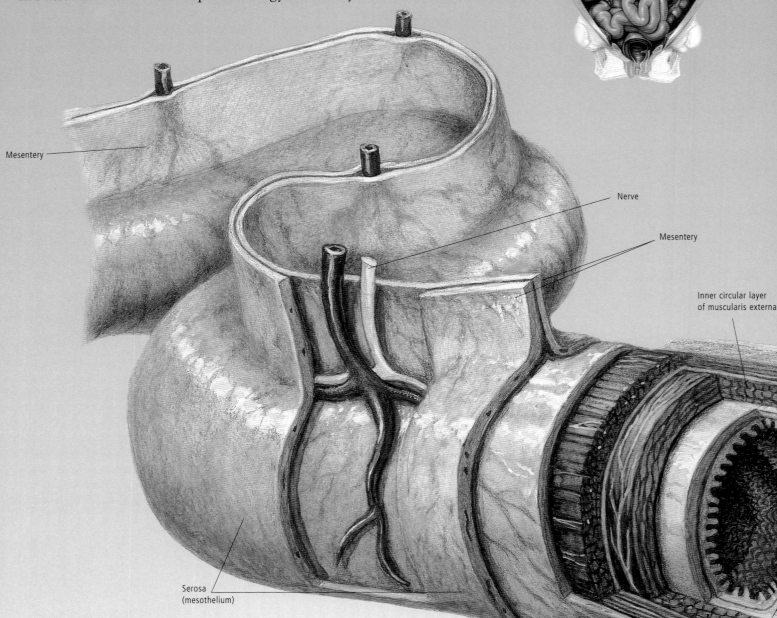

Mesentery

Nerve

Mesentery

Inner circular layer of muscularis externa

Serosa (mesothelium)

Serosa (connective tissue)

Plicae circulares

Intestinal jejunum cut-away

The jejunum has layers of smooth muscle running in circular and longitudinal directions, which work to push food along the intestinal tract. The folded inner lining of mucus membrane absorbs nutrients from the digested food as it passes through. The jejunum, as with the remainder of the intestines, is held in place by the mesentery. This membranous layer provides the blood supply, nerves and lymph vessels vital to the efficient functioning of the intestines.

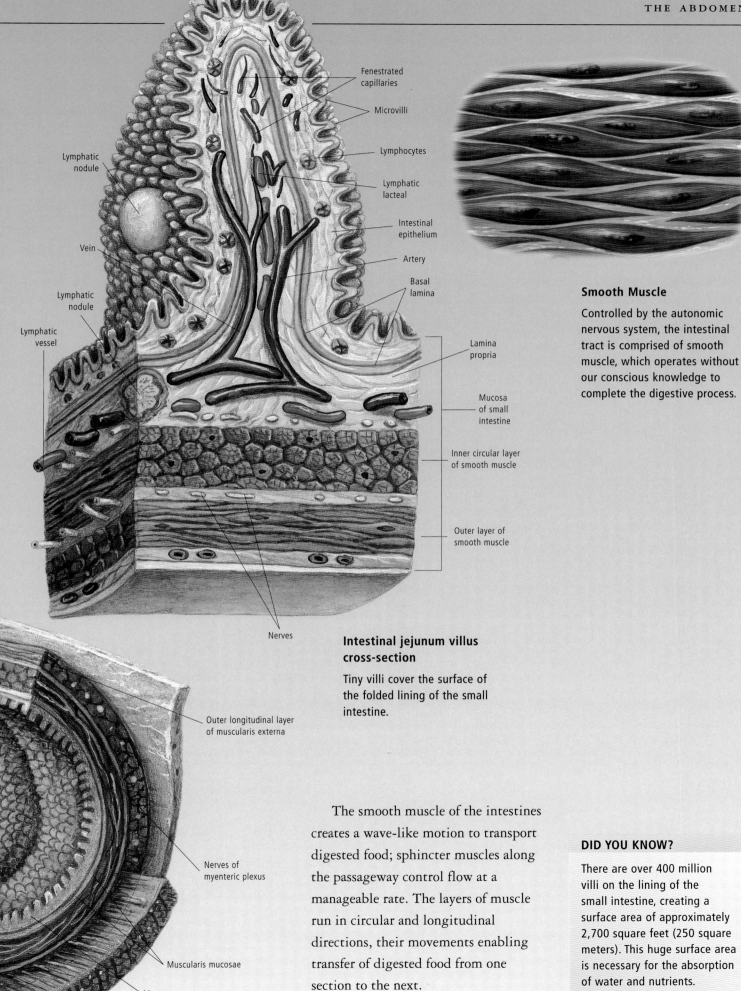

Fenestrated
capillaries

Microvilli

Lymphocytes

Lymphatic
lacteal

Intestinal
epithelium

Artery

Basal
lamina

Lamina
propria

Mucosa
of small
intestine

Inner circular layer
of smooth muscle

Outer layer of
smooth muscle

Lymphatic
nodule

Vein

Lymphatic
nodule

Lymphatic
vessel

Nerves

Outer longitudinal layer
of muscularis externa

Nerves of
myenteric plexus

Muscularis mucosae

Mucosa

Smooth Muscle

Controlled by the autonomic
nervous system, the intestinal
tract is comprised of smooth
muscle, which operates without
our conscious knowledge to
complete the digestive process.

Intestinal jejunum villus cross-section

Tiny villi cover the surface of
the folded lining of the small
intestine.

The smooth muscle of the intestines
creates a wave-like motion to transport
digested food; sphincter muscles along
the passageway control flow at a
manageable rate. The layers of muscle
run in circular and longitudinal
directions, their movements enabling
transfer of digested food from one
section to the next.

DID YOU KNOW?

There are over 400 million
villi on the lining of the
small intestine, creating a
surface area of approximately
2,700 square feet (250 square
meters). This huge surface area
is necessary for the absorption
of water and nutrients.

The Liver

Responsible for a variety of functions in the body, the liver is also the heaviest of all the body organs, weighing in at around $3^{1}/_{3}$ pounds (1.5 kilograms). Occupying the upper right side of the abdomen, the liver lies under the ribs and is attached to the diaphragm by folds of membrane. Separated into two lobes, the wedge-shaped liver is in contact with the gallbladder, the kidney, part of the duodenum, the esophagus, the stomach and part of the large bowel (large intestine). Folds of membrane attach the liver to the stomach and duodenum, while the gallbladder is usually attached to the liver by connective tissue.

The porta hepatis, the point of entry and exit for the blood vessels and ducts, lies on the visceral surface of the liver. Within the porta hepatis are the portal vein, the hepatic artery, and ducts of the biliary system.

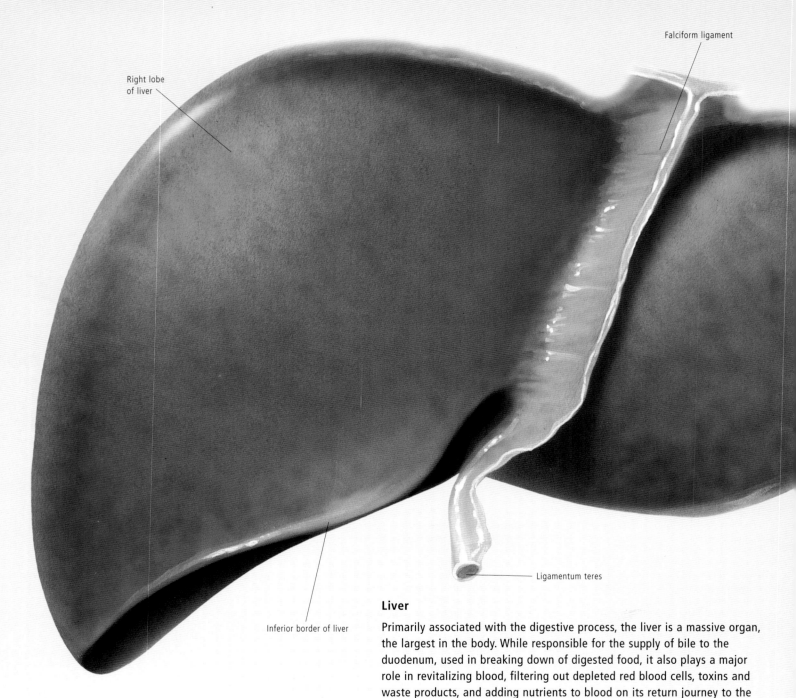

Falciform ligament

Right lobe of liver

Ligamentum teres

Inferior border of liver

Liver

Primarily associated with the digestive process, the liver is a massive organ, the largest in the body. While responsible for the supply of bile to the duodenum, used in breaking down of digested food, it also plays a major role in revitalizing blood, filtering out depleted red blood cells, toxins and waste products, and adding nutrients to blood on its return journey to the heart. In order to maintain the appropriate levels in the blood, the liver converts and stores glycogen, produces albumin and several clotting substances, and stores iron. The liver also makes and stores vitamin A.

Hepatic artery and Portal vein

The hepatic artery and portal vein are the major blood vessels supplying the liver. These vessels enter the liver at the porta hepatis, meaning 'door to the liver'.

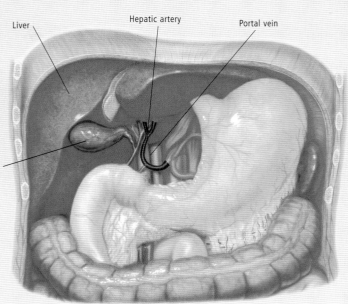

Liver

Hepatic artery

Portal vein

Gallbladder

NB: In this illustration the liver has been peeled back to show the position of the hepatic artery and the portal vein

Left lobe of liver

The liver is a multi-purpose organ, performing a variety of tasks essential to the efficient performance of the body. Blood from the gastrointestinal tract is returned to the liver, where the glucose content is converted into glycogen, which is then stored. Glycogen can be converted back to glucose when required, to maintain glucose levels in the blood. Two hormones released from the pancreas, insulin and glucagon, are important in the control of this function. The liver also serves as a filter for the blood, removing depleted red blood cells, excess nutrients, and removing and destroying toxins ingested with food and water.

The liver makes and stores vitamin A, which is essential for the well-being of the surface-lining tissues of the body, and stores iron, used in the production of hemoglobin in the blood. The liver also produces albumin, an important plasma protein in the blood, and several important substances involved in blood clotting.

Bile is produced in the liver for release into the duodenum to promote the breakdown of digested food. Bile arrives at the duodenum through the biliary system of ducts. Comprised of water, bile salts and a chemical called bilirubin, the bile is continually recycled, being extracted as it passes through the intestines and returned to the liver via the portal vein blood.

THE HEPATIC ARTERY

The hepatic artery supplies 30 percent of the liver's blood supply, the remainder coming from the portal vein. The common hepatic artery divides into right and left hepatic arteries to supply both sides of the liver with oxygenated blood.

THE PORTAL VEIN

Nutrient-rich blood is transported from the stomach, spleen, pancreas, gallbladder and intestines, back to the liver. This blood enters the portal vein on its way back to the liver. The portal vein is wide but short, entering the liver through the porta hepatis and ends as a network of capillaries throughout the liver called sinusoids. Depleted red blood cells are broken down and bacteria is removed.

The blood is replenished with nutrients, while any excess nutrients are removed and stored. The blood then leaves the liver through the hepatic veins which flow into the inferior vena cava.

THE LIVER LOBULES

The liver is made up of hexagonal lobules packed together with branches of the portal vein, hepatic artery and hepatic ducts at each corner of the hexagon. Running through the center of the lobule is a central vein. Each lobule is made up of layers of cells, each layer one cell thick. Venous blood returning from the intestines flows past the sheets of liver cells on its way to the central vein. Nutrients, bile salts, and toxic and waste substances are removed from the portal blood and processed. The central veins of all the lobules join together and contribute blood to the hepatic veins. These in turn drain into the inferior vena cava, which carries blood back to the heart.

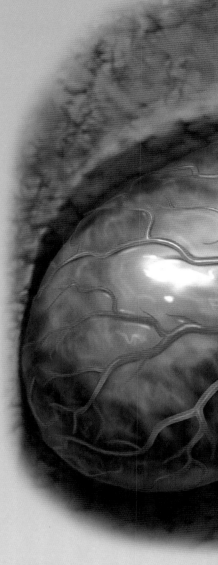

Bile duct

Artery

Collecting vein

Sublobular (intercalated) vein

Central vein

Interlobular bile duct

Branch of portal vein

Branch of hepatic artery

Opening of a liver sinusoid

Liver sinusoid

Gallbladder

The gallbladder is a small sac-shaped organ responsible for the storage and concentration of bile received from the liver. When prompted into action, the gallbladder releases bile into the bile duct, for transport to the duodenum where it assists in the digestive process.

Liver lobule

The liver is composed of hexagonal lobules, each with a central vein, and branches of the hepatic artery, portal vein and biliary vessels located at each corner of the hexagon. Each lobule is layer upon layer of cells, laid one on top of the other. The central vein drains blood from the lobule to the hepatic veins that lead out of the liver. Blood from the gut travels to the portal vein and on through its branches, before being returned to the heart. More blood is brought to the liver by the hepatic artery and its branches. Bile produced by the liver cells passes through the biliary duct system to the duodenum.

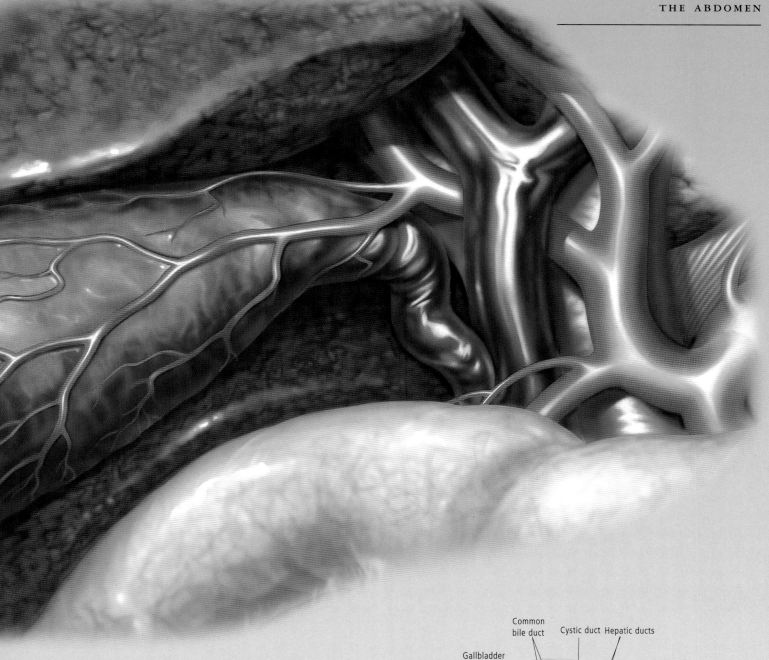

Common
bile duct   Cystic duct Hepatic ducts

Gallbladder

The Gallbladder

The gallbladder is a storage facility for bile produced by the liver.
A small sac-shaped organ tucked away under the liver, the gallbladder
forms part of the biliary system, joined to the bile duct by the cystic
duct. Bile produced by the liver cells flows along the bile ducts to the
gallbladder, where it is stored and concentrated prior to release into
the duodenum. The bile duct passes from the junction with the cystic
duct down through the head of the pancreas, joining with the main
duct of the pancreas, then continuing on to the duodenum, where the
bile is released. The bile serves to break down large globules of fat into
manageable particles, thereby improving their digestion and absorption.
Once this has been achieved, the bile is absorbed by the small intestine
and returned to the liver via the portal system of veins, to begin its cycle
once more. This process is known as enterohepatic biliary circulation.

Bile ducts

Bile, a fluid containing pigments (bilirubin), lecithin
and bile salts, is transported from the liver to the
gallbladder through the bile ducts, ready for release
into the duodenum.

The Pancreas

The pancreas is a member of two body systems: the digestive system and the endocrine system. It is responsible for supplying enzymes to the duodenum for use in the digestive process, while its endocrine function is to produce hormones.

Lying behind the stomach and in front of the large blood vessels (the aorta and inferior vena cava) running through the abdomen, the pancreas is a pennant-shaped gland, with a broad head tapering to a narrow tail. The head is framed by the duodenum, while the tail meets with the spleen.

Pancreas

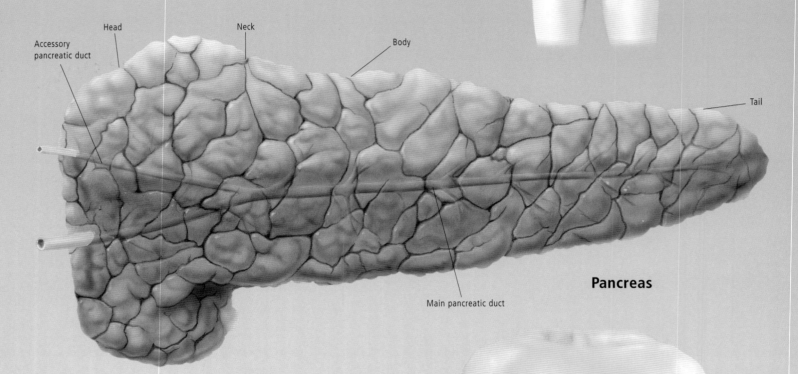

Head

Neck

Body

Accessory pancreatic duct

Tail

Pancreas

Main pancreatic duct

Comprised of both exocrine and endocrine cells to serve its dual purpose, the majority of the cells are exocrine cells supplying the digestive system with enzymes to aid in the digestion of food. A network of ducts convey the enzymes produced by the pancreas to the duodenum for use in the digestive process. Several protective measures are in place to prevent the enzymes from destroying the pancreas itself. These include measures such as storage of the enzymes in separate compartments, and the production of chemical inhibitors.

Clusters of endocrine cells are scattered throughout the pancreas, producing hormones used to regulate the body's metabolic activities. These hormones, insulin and glucagon, maintain and regulate blood sugar levels, and are produced in clusters of special cell types known as the islets of Langerhans.

Pancreas

The Adrenal Glands

The kidneys and adrenals are located at the back of the abdomen, with the adrenal glands sitting on top of the kidneys. The two adrenal glands are responsible for the production and release of hormones.

Rather triangular in shape and yellow brown in color, the adrenal glands have two layers: the outer layer or adrenal cortex, and the inner core, or medulla. The layers have differing structures, and produce and release different hormones. Hormones produced by the adrenal cortex include glucocorticoids, mineralocorticoids and sex steroids. These hormones play a role in maintaining fluid volume, blood pressure, blood volume and heart output, and also control sodium absorption. Androgens produced by the cortex contribute to the development of male sexual characteristics. Hormones produced by the medulla include epinephrine and nor-ephinephrine (adrenaline and noradrenaline). Secreted in response to stress, these two hormones act to raise the heart rate, increase blood supply to the muscles, increase the blood sugar levels and widen the airways, thereby preparing the body for stressful situations. Their action is called the "fight or flight" response, since they are the two options presented in situations of stress.

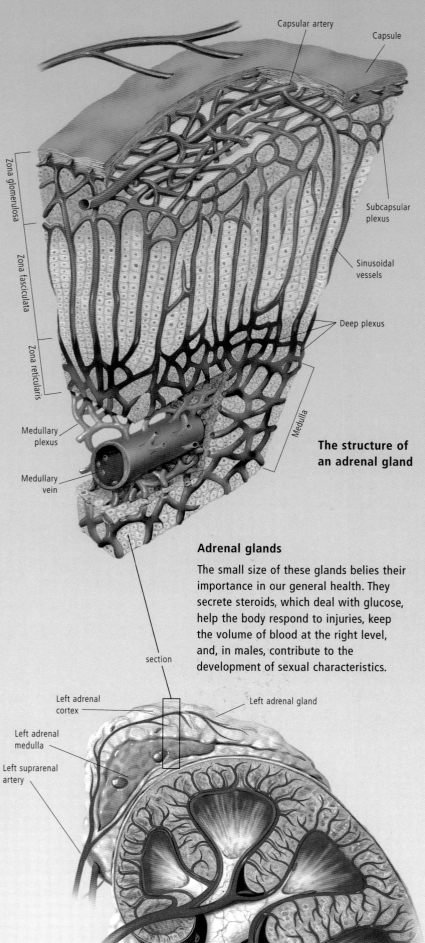

The structure of an adrenal gland

Capsular artery

Capsule

Subcapsular plexus

Sinusoidal vessels

Deep plexus

Zona glomerulosa

Zona fasciculata

Zona reticularis

Medulla

Medullary plexus

Medullary vein

Adrenal glands

The small size of these glands belies their importance in our general health. They secrete steroids, which deal with glucose, help the body respond to injuries, keep the volume of blood at the right level, and, in males, contribute to the development of sexual characteristics.

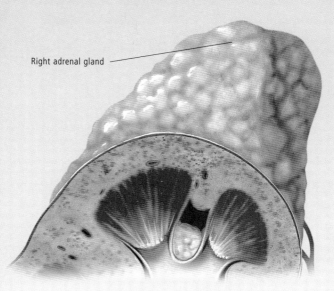

Right adrenal gland

section

Left adrenal cortex

Left adrenal medulla

Left suprarenal artery

Left adrenal gland

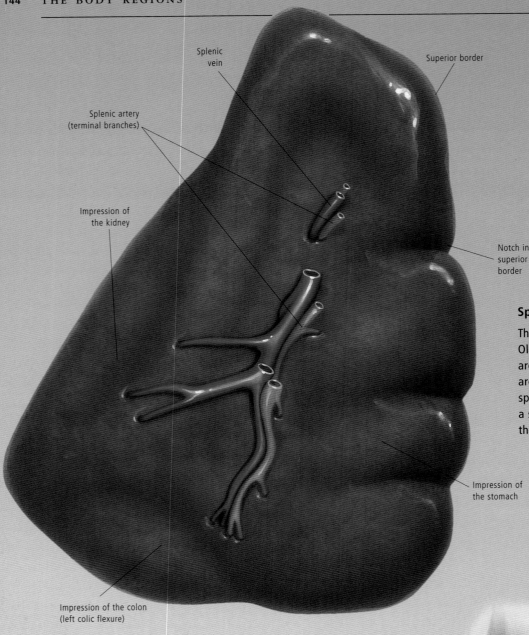

Splenic vein

Superior border

Splenic artery (terminal branches)

Impression of the kidney

Notch in superior border

Impression of the stomach

Impression of the colon (left colic flexure)

Spleen

The spleen is a filter for the blood. Old and abnormal red blood cells are broken down in the red pulp area. The white pulp area of the spleen is lymphatic tissue, providing a storehouse for lymphocytes of the immune system.

The Spleen

The spleen lies on the upper left side of the abdomen, beneath the diaphragm, and protected by the ninth, tenth and eleventh ribs. Similar in size to the heart, the spleen is the largest concentration of lymphatic tissue in the body, producing and storing lymphocytes—white blood cells crucial to immune response.

The spleen is a soft organ, red in color and pulpy in texture. This soft pulpy composition results in the surrounding organs of the kidneys, colon and stomach leaving an impression on its surface.

A capsular layer encases the red and white pulp that make up the spleen. The red pulp of the spleen filters blood through channels called sinuses, removing aged and abnormal red blood cells from the system. It is the role of the red pulp to form blood cells during fetal life. The white pulp of the spleen surrounds the splenic arteries. Lymphocytes, vital to the immune system, are stored in the white pulp.

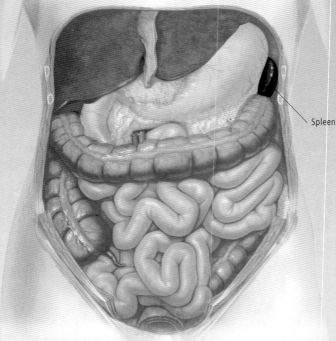

Spleen

The Groin and Abdominal Wall

The junction of the abdomen and the thigh is known as the groin or inguinal region. A concentration of lymph nodes serving the lower limbs is found in the groin region.

The Solar Plexus

Situated just below the diaphragm and behind the stomach, the solar plexus is a dense network of nerves on the abdominal aorta. These nerves branch out to all the abdominal organs and not only control these organs but also bodily functions such as intestinal contractions and adrenal secretions.

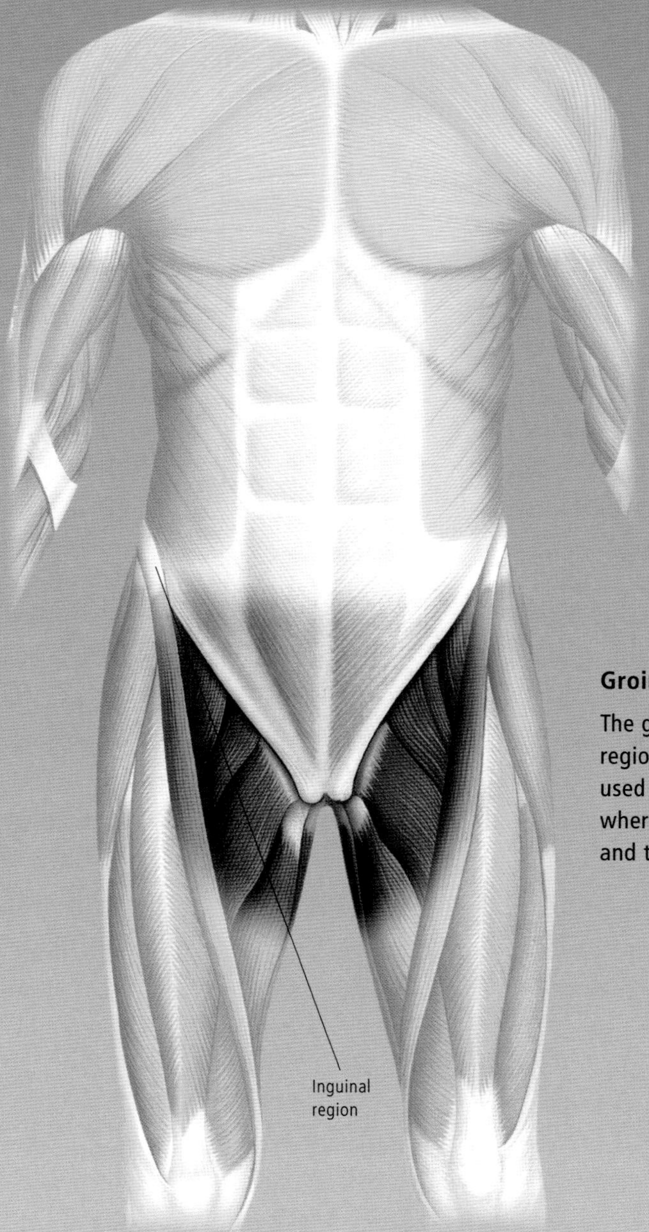

Groin

The groin, or inguinal region, is the term used for the region where the abdomen and thigh meet.

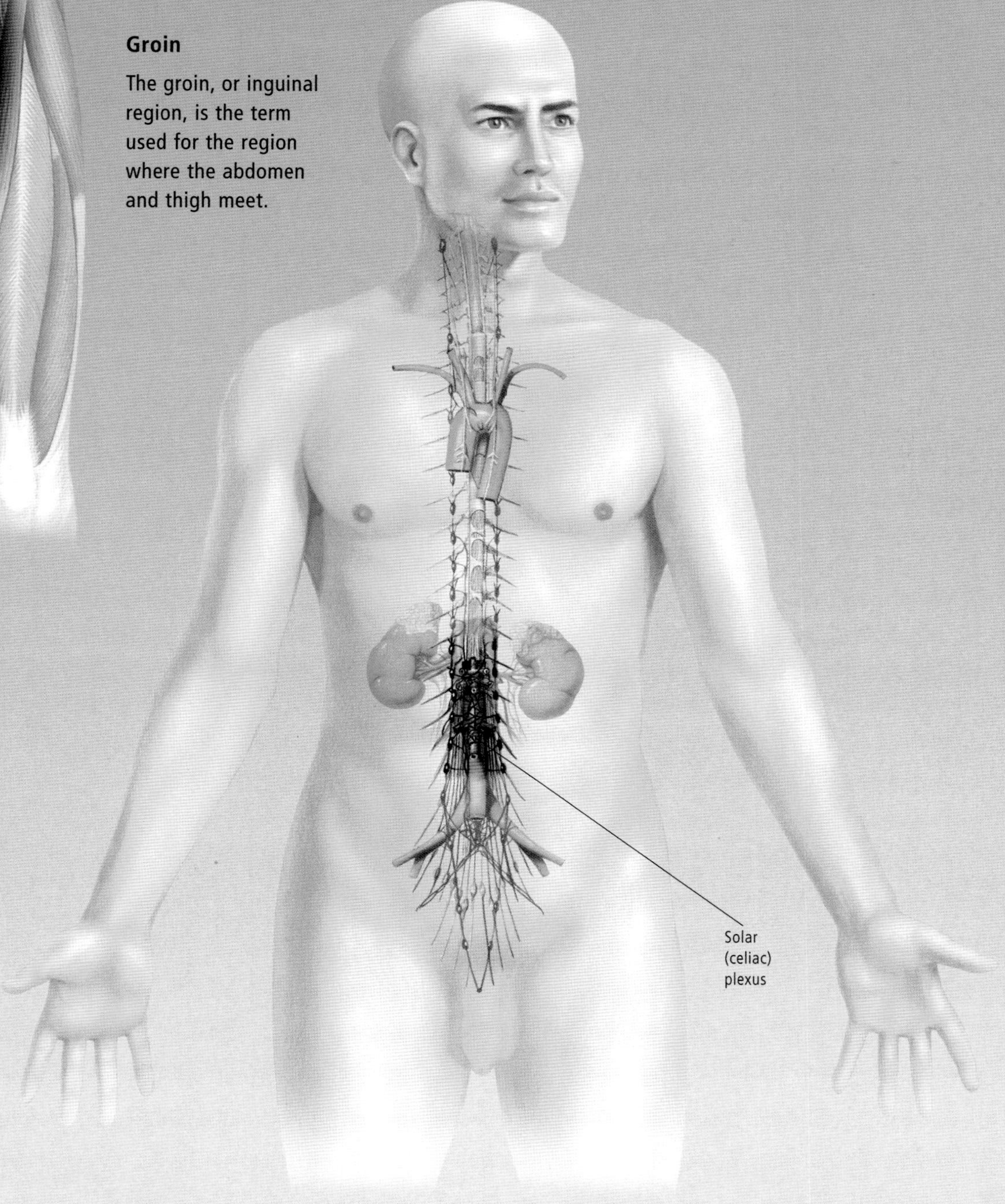

Inguinal region

Solar plexus

Also known as the celiac plexus, due to its location around the celiac artery, and "the pit of the stomach", the solar plexus is a network of nerves controlled by the autonomic nervous system. These nerves branch out to the abdominal organs, ensuring vital bodily functions remain operational.

Solar (celiac) plexus

The Urinary & Reproductive Organs

The Urinary Tract

Situated at the back of the abdomen, the urinary tract comprises the kidneys, ureters, bladder and urethra. The kidneys filter out waste products and excess water from the bloodstream in the form of urine. This urine travels along tubes called ureters to the bladder. From the bladder the urine passes out of the body through another tube called the urethra.

Male urinary system

The male urinary tract consists of the kidneys, ureters, bladder and urethra. The urethra serves a dual purpose, being a passageway for both urine and sperm.

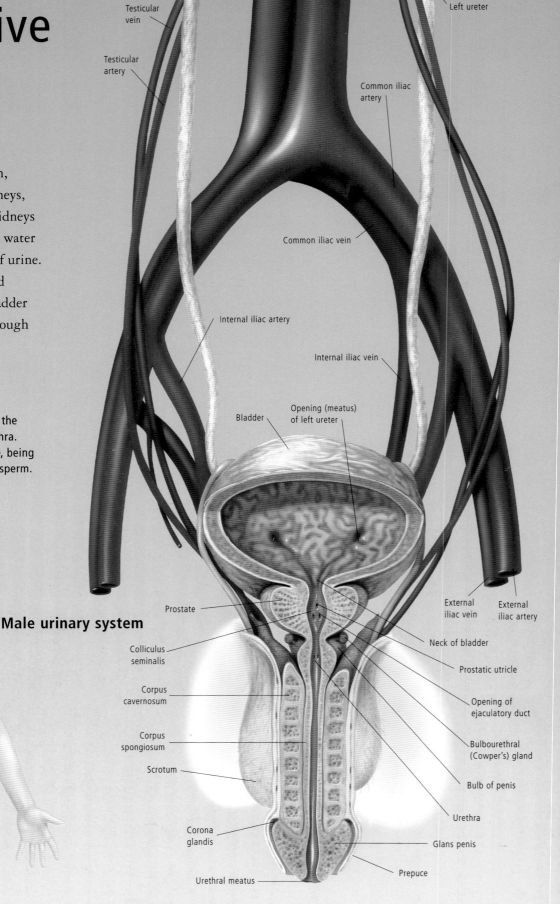

Male urinary system

Inferior vena cava

Abdominal aorta

Testicular vein

Left ureter

Testicular artery

Common iliac artery

Common iliac vein

Internal iliac artery

Internal iliac vein

Bladder

Opening (meatus) of left ureter

External iliac vein

External iliac artery

Prostate

Neck of bladder

Colliculus seminalis

Prostatic utricle

Corpus cavernosum

Opening of ejaculatory duct

Corpus spongiosum

Bulbourethral (Cowper's) gland

Scrotum

Bulb of penis

Urethra

Corona glandis

Glans penis

Prepuce

Urethral meatus

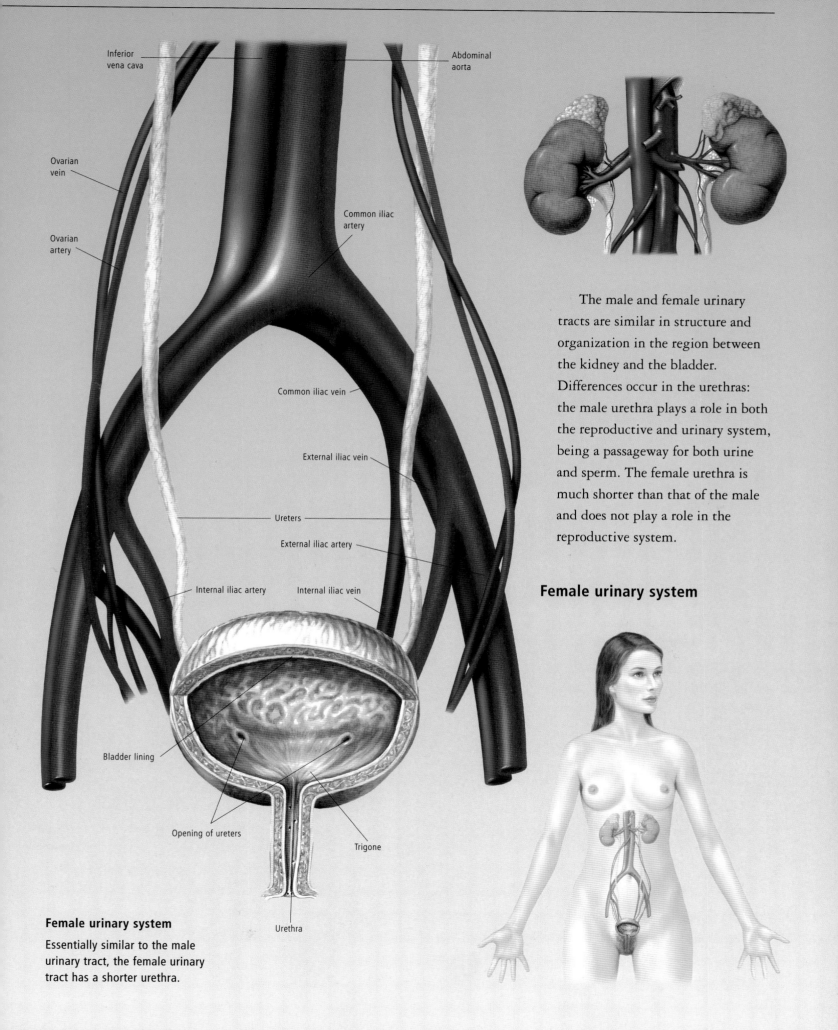

Inferior vena cava

Abdominal aorta

Ovarian vein

Ovarian artery

Common iliac artery

Common iliac vein

External iliac vein

Ureters

External iliac artery

Internal iliac artery

Internal iliac vein

Bladder lining

Opening of ureters

Trigone

Urethra

The male and female urinary tracts are similar in structure and organization in the region between the kidney and the bladder. Differences occur in the urethras: the male urethra plays a role in both the reproductive and urinary system, being a passageway for both urine and sperm. The female urethra is much shorter than that of the male and does not play a role in the reproductive system.

Female urinary system

Female urinary system

Essentially similar to the male urinary tract, the female urinary tract has a shorter urethra.

Kidneys

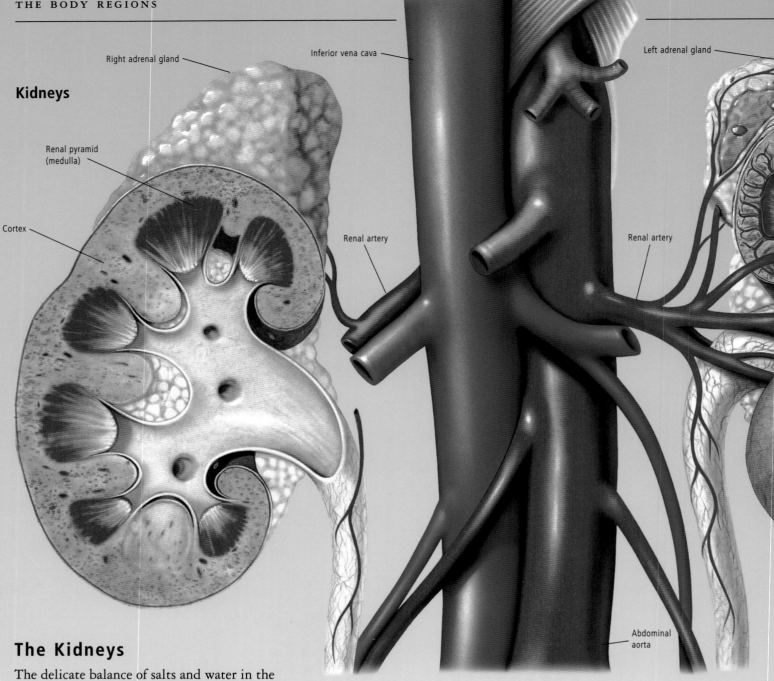

Right adrenal gland

Inferior vena cava

Left adrenal gland

Renal pyramid
(medulla)

Cortex

Renal artery

Renal artery

Abdominal
aorta

The Kidneys

The delicate balance of salts and water in the
body is controlled by the kidneys. Connected to
the back wall of the abdomen by a layer of connective tissue, the two kidneys sit either side of the spine, and
are topped by the adrenal glands. At the center of each
kidney is an indentation known as the renal hilus, the exit
point for the ureter and and entry and exit point for nerves,
blood and lymphatic vessels; the ureters join with the
kidneys at the renal pelvis. Each kidney has two layers,
an outer layer, the cortex, and an inner layer, the medulla.
The medulla contains between 8 and 18 renal pyramids.
These pyramids are formed with their apex pointing
towards the renal hilus. The cortex fills the space between
each pyramid, known as a renal column, and forms the
outer layer around the entire kidney. Each kidney is then
wrapped in a protective membrane, the renal capsule.

The function of the kidneys is to filter the blood, removing waste matter and excess water; these are then excreted as
waste in the form of urine. The kidneys are also responsible
for regulating and maintaining the salt and water levels in
the body. The kidneys respond to fluctuating levels by returning more water to the body when the body needs to conserve water or dilute salts. When the body has excess levels
of water, the kidneys excrete the excess through the urinary
tract, thus ensuring the correct levels of water and salts in
the body are constantly maintained.

The kidney also performs an endocrine function, releasing
the hormone erythropoietin into the blood; this hormone
plays a role in blood formation.

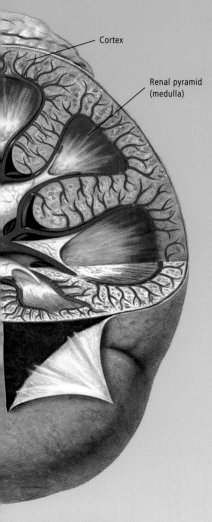

Cortex

Renal pyramid
(medulla)

THE RENAL ARTERIES

Running between the two kidneys are the major blood vessels of the aorta and the inferior vena cava. Leading off either side of the abdominal aorta are the two large blood vessels that supply the kidneys, the renal arteries. These arteries enter the kidney at the hilus, branching off to the adrenal glands and ureters, before dividing into two large branches, which are called the anterior and posterior divisions of each artery. These two branches then subdivide until they form the capillaries that supply oxygen to the kidney tissue and filter the blood.

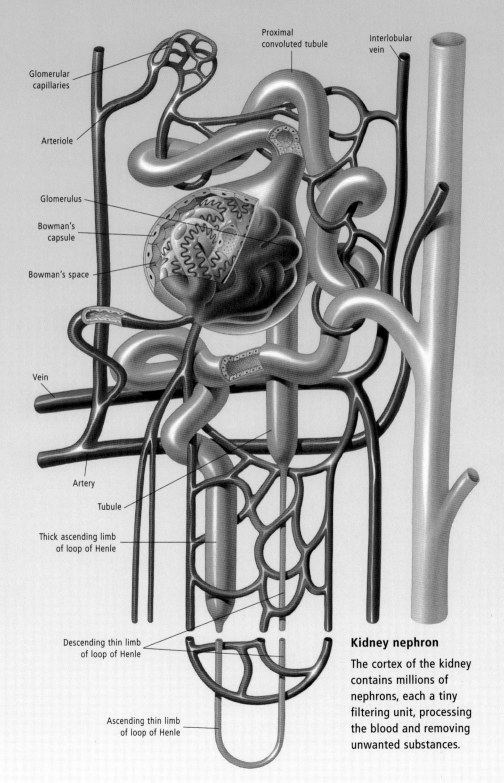

Proximal
convoluted tubule

Interlobular
vein

Glomerular
capillaries

Arteriole

Glomerulus

Bowman's
capsule

Bowman's space

Vein

Artery

Tubule

Thick ascending limb
of loop of Henle

Descending thin limb
of loop of Henle

Ascending thin limb
of loop of Henle

THE NEPHRON

Over a million nephrons in each kidney function to filter the blood. Each nephron spans between the two layers of the kidney. Lying in the cortex is the glomerulus, a ball-shaped capillary network encapsulated by a membrane known as Bowman's capsule. Together, these structures filter the blood, producing a liquid (filtrate) containing minerals, waste and water. The purified blood is returned to the body, while the filtrate passes into the renal tubule, which runs through a renal pyramid in the medulla. Any useful components of the filtrate are reabsorbed by the capillaries of the renal tubule—resulting in almost 99 percent of filtrate being returned to the general circulation. The remainder, along with any waste secreted by the renal tubule capillaries, is excreted as urine.

Kidney nephron

The cortex of the kidney contains millions of nephrons, each a tiny filtering unit, processing the blood and removing unwanted substances.

The Bladder

Lying in the pelvic cavity, the bladder is a storage area for urine. Urine excreted by the kidneys travels along the thin muscular tubes of the ureters; its passage is controlled by the muscular contractions and relaxations of smooth muscle. The bladder is a muscular sac, with the ureters entering from above and the urethra joining at the base; these three openings form the three corners of the area known as the trigone. Once in the bladder, urine is stored until the bladder fills. The capacity of the adult bladder is about 1 pint (475 milliliters), and around 24–68 fluid ounces (700–2000 milliliters) of urine are expelled daily. When the bladder is full, the neck of the bladder relaxes to release urine into the urethra. This normal reflex can be suppressed until there is an opportunity to relieve the bladder. Urine is then expelled by the bladder via the urethra.

The Urethra

Under our conscious control, the bladder muscles release urine into the urethra. The urethra is the final passageway for urine before leaving the body. The male and female urethra differ, with the female urethra being relatively short. It passes directly through the pelvic floor and opens out in front of the entrance to the vagina.

The male urethra is longer, usually 8 inches (20 centimeters) long, and plays a role in both the urinary and reproductive systems, providing a common passageway for both sperm and urine. The prostate gland lies immediately below the bladder, encircling the urethra. The urethra extends from the bladder, through the prostate gland, where it is joined by sperm ducts, and continues to the tip of the penis.

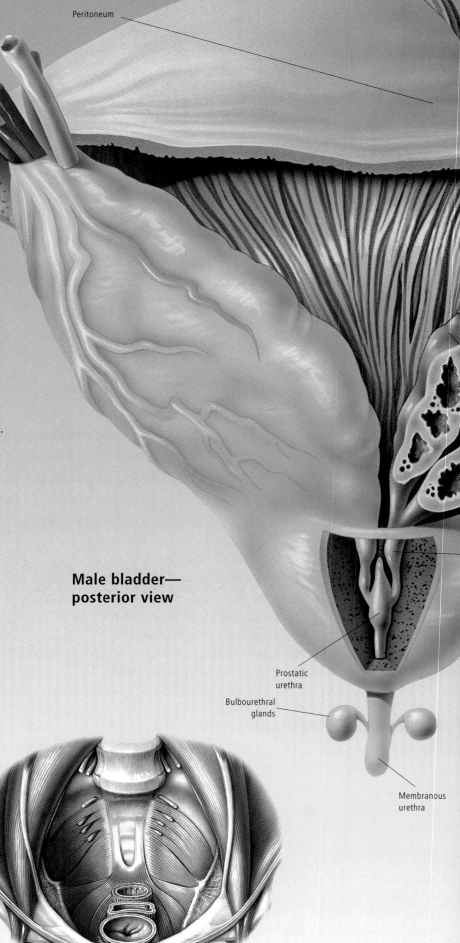

Peritoneum

Male bladder— posterior view

Prostatic urethra

Bulbourethral glands

Membranous urethra

Pelvic floor muscles

The pelvic floor muscles support the pelvic organs, including the bladder, and the prostate in men. There are several openings in the pelvic floor, including the opening for the urethra to the outside of the body.

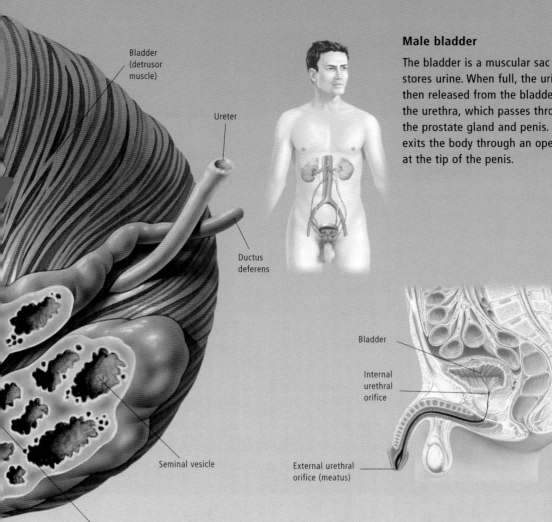

Bladder (detrusor muscle)

Ureter

Ductus deferens

Seminal vesicle

Ampulla of ductus deferens

Ejaculatory duct

Prostate

Male bladder

The bladder is a muscular sac which stores urine. When full, the urine is then released from the bladder into the urethra, which passes through the prostate gland and penis. Urine exits the body through an opening at the tip of the penis.

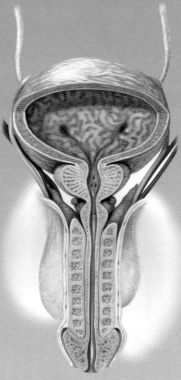

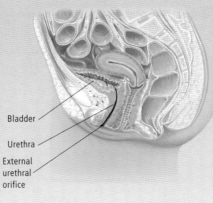

Bladder

Internal urethral orifice

External urethral orifice (meatus)

Male urethra

The male urethra is a passageway for both urine and sperm. It connects both the bladder and the sperm ducts from the prostate gland to the outside of the body.

Sphincter muscle

Sphincter muscles in the urinary tract control the release of urine. When these circular muscles relax, urine is allowed to pass through the opening; when the muscles contract, the opening is effectively sealed off.

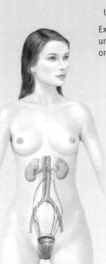

Bladder

Urethra

External urethral orifice

Female urethra

The female urethra is relatively short. It exits the body through the pelvic floor, through an opening immediately in front of the vagina.

Female bladder

The female bladder is similar to the male bladder, but the adjoining urethra is much shorter. The uterus lies immediately above the bladder.

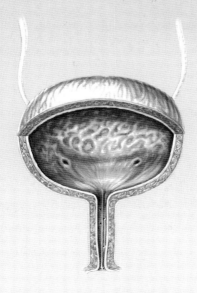

The Male Reproductive Organs

The male reproductive organs include the testes, the epididymis, the prostate gland and the penis.

THE PROSTATE GLAND

This small gland surrounds the neck of the bladder and encircles the urethra. Composed of muscular and glandular tissue, the prostate is shaped like an inverted pyramid. Containing two major groups of glands, the prostate is joined to the urethra by ducts, which carry its glandular secretions to combine with sperm. The secretions from the prostate and seminal vesicles contain glucose and enzymes which provide the energy spermatozoa need to propel toward the ovum. About a quarter of the seminal fluid content is supplied by the prostate.

THE PENIS

The penis is part of both the urinary and reproductive systems, providing an outlet for both semen and urine. Comprised of two cylinders of sponge-like tissue and a third cylinder containing the urethra; all three cylinders are encased in a layer of connective tissue. The urethra ends in an external swelling at the tip of the penis, the glans. The glans is particularly sensitive and, in an uncircumcised penis, is covered by a protective foreskin (prepuce).

The interconnecting spaces between the cylinders are filled with blood. The mechanism of erection is controlled by the autonomic nervous system. When the blood flow out of the interconnecting spaces of the cylinders is halted by closure of the veins, the cylinders become engorged with blood, thereby causing an erection. The erection ceases when the veins open, allowing blood to flow back into the body's circulation.

Male reproductive organs

Sperm is produced in the testes, then passes along the ductus deferens which joins the duct of the seminal vesicle to form the ejaculatory duct. Secretions from the prostate and seminal vesicles combine with the sperm during ejaculation to form the seminal fluid.

Penis

Testis

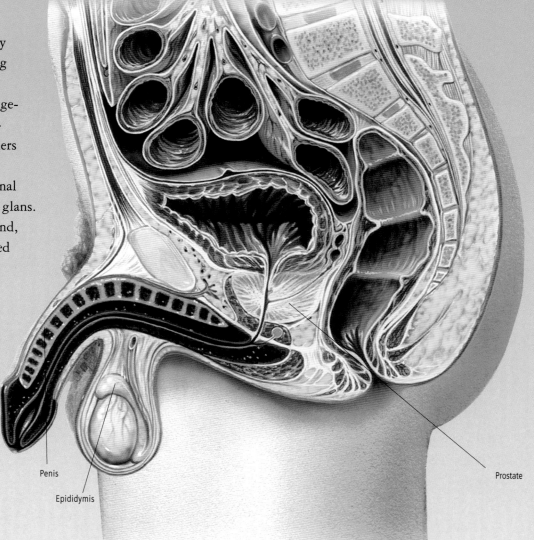

Penis

Epididymis

Prostate

Penis

The penis is the external male reproductive and urinary organ, providing an outlet for both urine and semen. The penis is attached to the pelvic bone by connective tissue and is comprised of three layers of erectile tissue with an outer layer of connective tissue. When stimulated, the penis becomes engorged with blood, resulting in an erection.

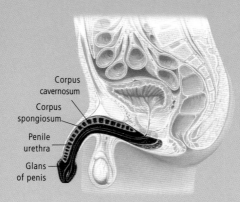

Corpus cavernosum

Corpus spongiosum

Penile urethra

Glans of penis

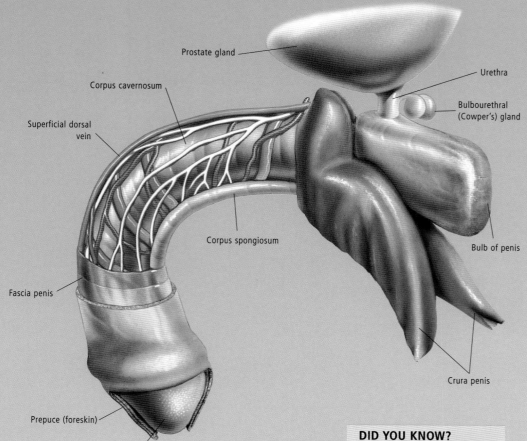

Prostate gland

Corpus cavernosum

Superficial dorsal vein

Urethra

Bulbourethral (Cowper's) gland

Corpus spongiosum

Bulb of penis

Fascia penis

Crura penis

Prepuce (foreskin)

Glans of penis

DID YOU KNOW?

Sperm production begins at the onset of male puberty and continues throughout life. Every day, several hundred million sperm are produced.

Prostate gland

The prostate is made up of glandular and muscular tissue, and sits beneath the bladder. It secretes some of the fluid present in semen.

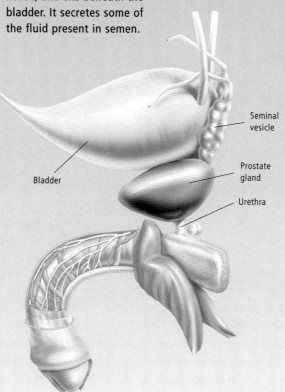

Bladder

Seminal vesicle

Prostate gland

Urethra

Prostate–urinary link

The prostate gland surrounds the neck of the urethra. Because of its location, enlargement of the prostate (which may occur with age or disease) causes difficulty in urination.

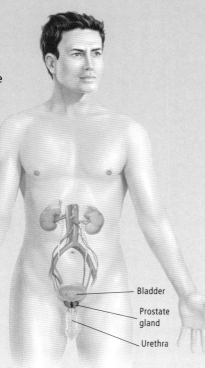

Bladder

Prostate gland

Urethra

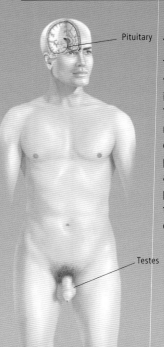

Pituitary

Testes

Testosterone

Triggered by hormones in the pituitary, testosterone is secreted by the testes. Testosterone controls the many physical changes experienced during male puberty, including facial and pubic hair, sperm production, deepening of the voice and enlargement of the genitals.

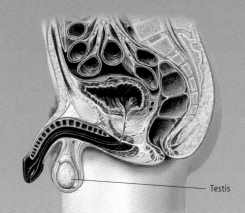

Testis

Sperm

With an appearance similar to tadpoles, sperm even use a "swimming" motion to propel forward.

Tail

Testes

The main organs of the male reproductive system, the testes are contained in the scrotum, situated outside the body, at a temperature lower than that of the body, to create the optimal environment for sperm production.

THE SCROTUM

Attached to the male perineum, the scrotum is a wrinkled bag of skin and soft tissue lying behind the penis that holds the testes and the lower part of the spermatic cords. Its wrinkled appearance is due to the contractions of a thin muscular layer beneath the skin.

THE TESTES

Contained in the scrotum, the testes, or testicles, are the major organs of reproduction in the male. These reproductive organs become fully functional at puberty, when triggered by the pituitary gland to commence hormone and sperm production. Once sperm production begins, it continues to old age. Sperm cannot be properly developed at body temperature, but because the testes are located outside the body, where the temperature is slightly lower, sperm production is optimized. Within the scrotum, the testes are wrapped in membrane, the tunica vaginalis, and the epididymis runs along the back of each of the testes. The epididymis has a broad head, tapering to a narrow tail, then folds back on itself to become the ductus deferens, or vas deferens.

The testes are divided into many compartments, called lobules, and these lobules each contain seminiferous tubules that produce sperm. These tubules join together at ducts which lead to the epididymis.

Cremaster muscle and fascia

Ductus deferens

Epididymis

Testis

Scrotal skin

Testes

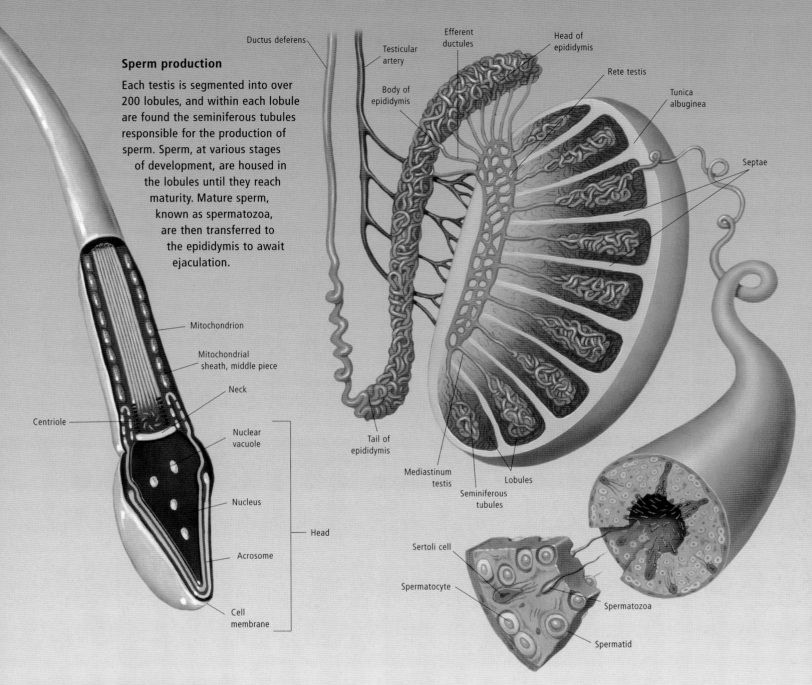

Sperm production

Each testis is segmented into over 200 lobules, and within each lobule are found the seminiferous tubules responsible for the production of sperm. Sperm, at various stages of development, are housed in the lobules until they reach maturity. Mature sperm, known as spermatozoa, are then transferred to the epididymis to await ejaculation.

Ductus deferens

Testicular artery

Body of epididymis

Efferent ductules

Head of epididymis

Rete testis

Tunica albuginea

Septae

Mitochondrion

Mitochondrial sheath, middle piece

Neck

Centriole

Nuclear vacuole

Nucleus

Head

Acrosome

Tail of epididymis

Cell membrane

Mediastinum testis

Lobules

Seminiferous tubules

Sertoli cell

Spermatocyte

Spermatozoa

Spermatid

SPERM AND SPERM PRODUCTION

Sperm are produced by the seminiferous tubules, with each tubule containing sperm at different stages of development. When they reach the final stage of development, spermatozoa, they move through the testes and make their way to the epididymis. Here, they mature and await ejaculation, brought about by muscular contractions. The sperm are propelled into the urethra where they combine with seminal fluid from the prostate and seminal vesicles; these seminal fluids react with the sperm to mobilize them. Semen contains 90 percent seminal fluid and 10 percent sperm and epididymal fluid.

Ejaculation releases the semen, which contains between 80 million and 300 million sperm. Sperm have a tadpole-like appearance, with the head containing the chromosomes that will determine the sex of a fertilized ovum, and enzymes to aid its penetration of the coating of the ovum. The tail of the sperm propels it, in a "swimming" action, on its quest to meet with the ovum. Despite the high number of sperm in the semen, usually only one sperm will fertilize the ovum.

TESTOSTERONE

The hormone testosterone is produced by the testes and is responsible for triggering the physical changes during male puberty. Testosterone stimulates growth of facial and pubic hair, enlargement of the larynx and deepening of the voice, enlargement of the penis and testes, and an increase in muscular strength.

The Female Reproductive Organs

The female organs of reproduction consist of the ovaries, the fallopian tubes, the uterus and the vagina.

THE UTERUS

The uterus consists of the top (fundus), body and cervix. Joining with the top of the uterus and extending out from each side are the fallopian tubes. At the end of the fallopian tubes are fimbriae, feathery finger-like projections, whose role is to catch the ovum released by the nearby ovaries. The uterus is connected to the vagina below.

Held in place in the pelvis by the broad ligament, with further support provided by ligaments attaching the cervix to the wall of the pelvis, and cradled by the pelvic floor muscles, the uterus, or womb, is the organ of gestation. Usually pear-shaped, the uterus undergoes enormous expansion during pregnancy. The organs around the uterus must make room, by compressing or moving, to accommodate the growing fetus within the womb.

The uterus is comprised of several layers: the inner layer is the endometrium, the middle layer is the myometrium and the outer layer is the peritoneum. The endometrium is a glandular layer that undergoes cyclical changes in thickness and composition during the 28-day menstrual cycle. Preparing its surface in readiness for a fertilized ovum, the endometrium breaks down if the ovum remains unfertilized. This breakdown material and the subsequent sloughing of the endometrial surface is expelled from the cervix into the vagina as menstrual fluid.

THE CERVIX

The cervix, a cylindrical muscular tube, is the lower part of the uterus, with the cervical canal joining the body of the uterus with the vagina. Sperm passes through the cervix on its way to the uterine body, and menstrual blood is expelled through the cervix to its exit point, the vagina. The cervix has a mucus-secreting lining, which protects it from sloughing away when the endometrium breaks down and is shed. During pregnancy, this mucus forms a protective "plug", effectively sealing off the cervix.

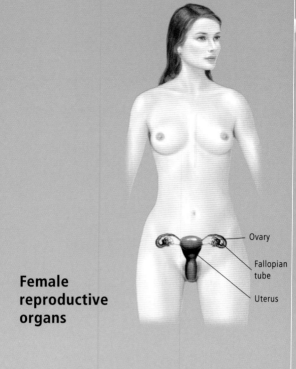

Female reproductive organs

Ovary
Fallopian tube
Uterus

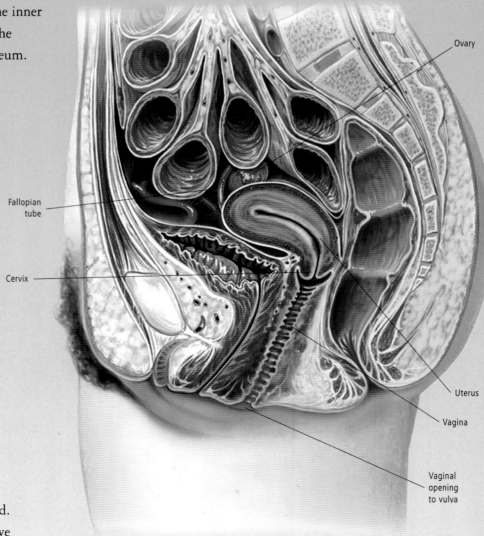

Ovary
Fallopian tube
Cervix
Uterus
Vagina
Vaginal opening to vulva

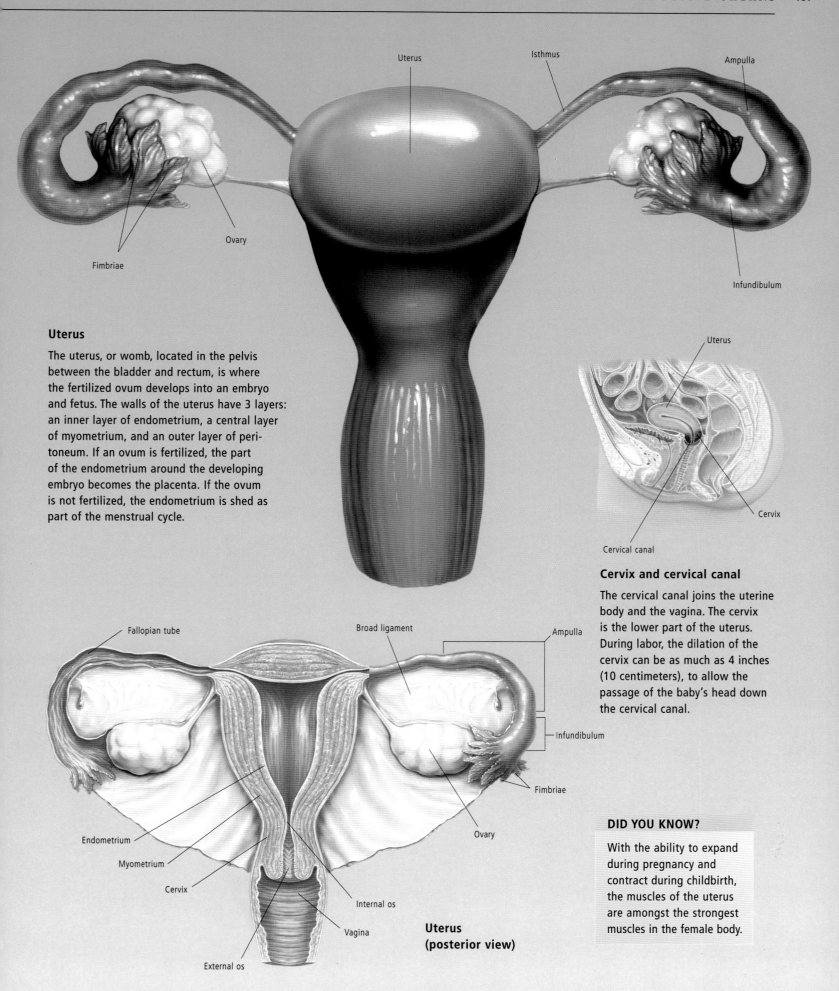

Uterus

Isthmus

Ampulla

Ovary

Fimbriae

Infundibulum

Uterus

The uterus, or womb, located in the pelvis between the bladder and rectum, is where the fertilized ovum develops into an embryo and fetus. The walls of the uterus have 3 layers: an inner layer of endometrium, a central layer of myometrium, and an outer layer of peritoneum. If an ovum is fertilized, the part of the endometrium around the developing embryo becomes the placenta. If the ovum is not fertilized, the endometrium is shed as part of the menstrual cycle.

Uterus

Cervix

Cervical canal

Cervix and cervical canal

The cervical canal joins the uterine body and the vagina. The cervix is the lower part of the uterus. During labor, the dilation of the cervix can be as much as 4 inches (10 centimeters), to allow the passage of the baby's head down the cervical canal.

Fallopian tube

Broad ligament

Ampulla

Infundibulum

Fimbriae

Endometrium

Myometrium

Cervix

Internal os

Ovary

Vagina

External os

**Uterus
(posterior view)**

DID YOU KNOW?

With the ability to expand during pregnancy and contract during childbirth, the muscles of the uterus are amongst the strongest muscles in the female body.

THE OVARIES

Held in place by the broad ligament of the uterus, the ovaries are oval in shape, measuring about $1\frac{1}{2}$ inches across, and lie near the side walls of the pelvis. In women of reproductive age, the ovaries contain thousands of undeveloped follicles as well as several ova at various stages of development. The ova are contained in a sac-like structure, known as a follicle. Of these developing ova, usually only one will go on to be released at ovulation, with the remainder degenerating at various stages of maturity. The surviving follicle matures into a Graafian follicle. At ovulation, one of the ovaries releases the ovum, which is drawn into the nearby fallopian tube. The remains of the Graafian follicle develop into the corpus luteum, in preparation for fertilization of the ovum. If the ovum remains unfertilized, the corpus luteum breaks down just prior to the start of the next menstrual cycle.

THE FALLOPIAN TUBES

The fallopian tubes (or uterine tubes) are trumpet shaped, with their narrow end leading into the uterus, and the flared end situated close to the ovary. This flared end has tiny finger-like projections called fimbriae, which gather up the released ovum. The ovum then moves along the fallopian tube toward the uterus. If the ovum is fertilized it embeds itself in the wall of the uterus; if the ovum remains unfertilized, the endometrium breaks down and menstrual bleeding occurs.

THE MENSTRUAL CYCLE

The menstrual cycle is a 28-day cycle, usually considered to begin at the onset of menstruation, a process which lasts, on average, 5 days. The remainder of the cycle comprises 2 phases, the proliferative phase (the days prior to ovulation), and the secretory phase (the days following ovulation).

After menstruation (Days 1–6), the inner lining, the endometrium, must regenerate itself. It regenerates during the proliferative phase (Days 7–13), rebuilding the lining of the uterus in readiness for a fertilized ovum. Ovulation (Day 14) signals the release of a mature ovum into the fallopian tubes. During the secretory phase (Days 15–28) the ovum travels along the fallopian tube to the uterus. If the ovum is not fertilized, hormone levels drop and the blood vessels in the uterus constrict. The endometrium breaks down, and is discharged. The cycle then begins again.

Ovaries

These almond-shaped organs contain thousands of undeveloped follicles. The follicles develop, a few at a time, ready to be released at ovulation. Usually only one ovum is released, with the remainder of the developing ova degenerating at various stages of maturity.

Ovaries

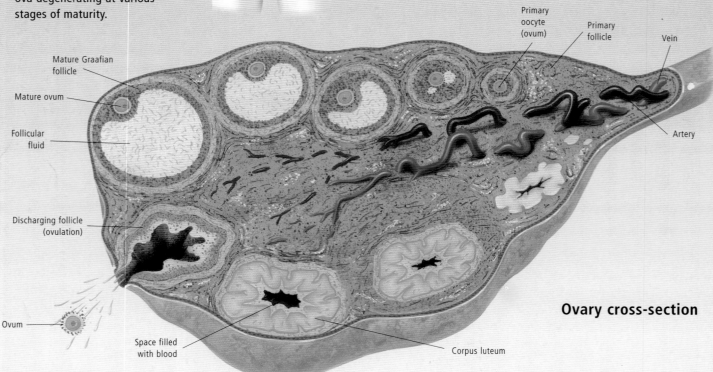

Primary oocyte (ovum)

Primary follicle

Vein

Mature Graafian follicle

Mature ovum

Follicular fluid

Artery

Discharging follicle (ovulation)

Ovum

Space filled with blood

Corpus luteum

Ovary cross-section

Menstrual cycle
The menstrual cycle occurs over a 28-day period.
(a) Days 1–6: Menstruation—the endometrium (lining of the uterus) breaks down and is discharged.
(b) Days 7–13: Proliferative phase—regeneration of the endometrium takes place.
(c) Day 14: Ovulation—the ovum is released into the fallopian tube by the ovary.
(d) Days 15–28: Secretory phase—after ovulation, the ovum moves along the fallopian tube to the uterus, the lining of which has been thickened by hormones in preparation for a fertilized ovum. However, if the ovum is not fertilized, hormone levels drop and blood vessels in the uterus constrict.
The menstrual cycle begins again when the unused endometrial blood and tissue is discharged from the body.

Menstrual cycle

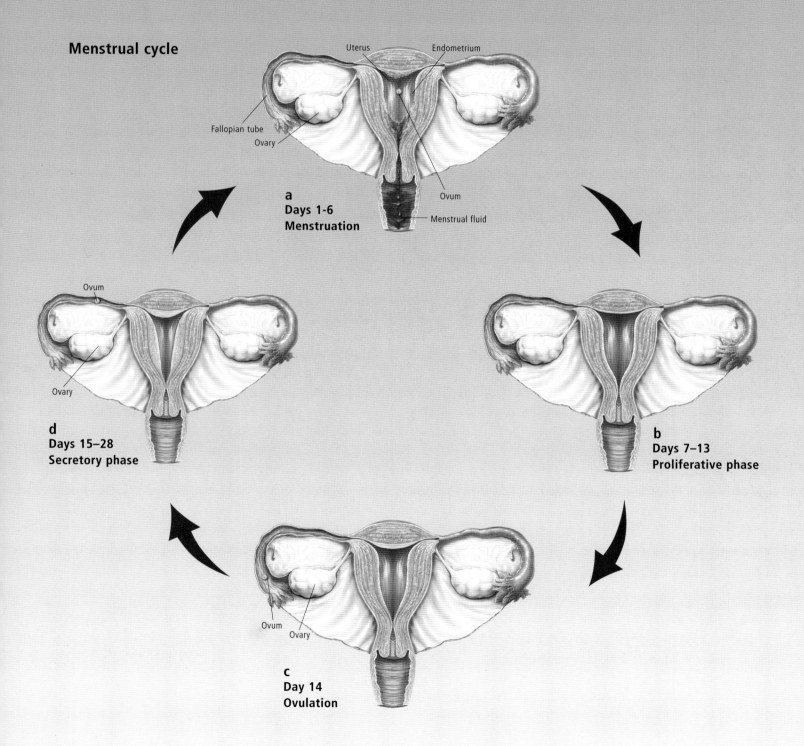

Uterus   Endometrium

Fallopian tube
Ovary

Ovum

Menstrual fluid

a
Days 1-6
Menstruation

Ovum

Ovary

d
Days 15–28
Secretory phase

b
Days 7–13
Proliferative phase

Ovum
Ovary

c
Day 14
Ovulation

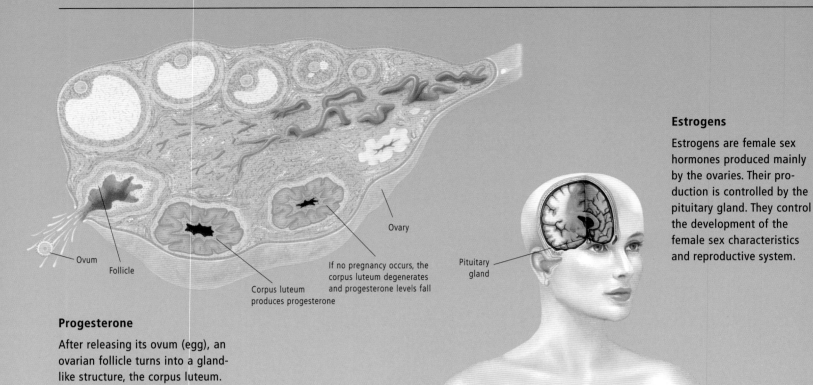

Ovum

Follicle

Corpus luteum
produces progesterone

If no pregnancy occurs, the
corpus luteum degenerates
and progesterone levels fall

Ovary

Pituitary
gland

Estrogens

Estrogens are female sex
hormones produced mainly
by the ovaries. Their production is controlled by the
pituitary gland. They control
the development of the
female sex characteristics
and reproductive system.

Progesterone

After releasing its ovum (egg), an
ovarian follicle turns into a gland-
like structure, the corpus luteum.
This produces progesterone which
prepares the uterus for pregnancy.

ESTROGENS

Estrogens are female sex hormones produced primarily
in the ovaries. Estrogens control the development
of female sex characteristics. Estrogens also trigger
the rebuilding of the endometrium in readiness
for a fertilized ovum. During pregnancy, high
levels of estrogen are secreted, then a few
days after childbirth, the levels drop.

PROGESTERONE

Progesterone is a sex hormone produced by
the corpus luteum, formed from the remains
of the Graafian follicle after the release of
the ovum. Progesterone stimulates the
endometrium to secrete a fluid which
protects and nourishes the fertilized
ovum in the uterus before implantation, until this role can be
taken over by the placenta at the
end of the first trimester. After
childbirth, the levels of progesterone
drop dramatically. If the ovum is not
fertilized, at around Day 26 of the menstrual cycle, the
corpus luteum ceases to function, causing a rapid decline
in progesterone development; this brings about changes
in the lining of the uterus which lead to menstruation.

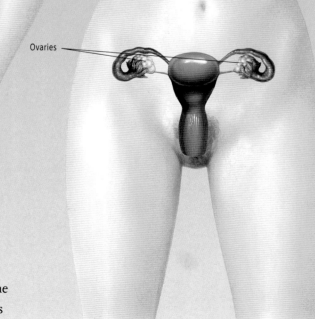

Ovaries

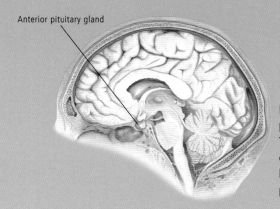

Anterior pituitary gland

Female hormone production
The anterior pituitary gland controls the menstrual hormones estrogen and progesterone.

DID YOU KNOW?

At birth, a female has about 2 million developing egg cells. By the time puberty is reached, the number has dropped to around 350,000, and by menopause there are none.

THE VAGINA

The vagina is a fibromuscular tube, joining to the cervix and thus connecting the uterus to the outside of the body. The tube consists of a muscular external wall and an inner mucosa, with an internal cavity. Normally the inner surfaces lie close together, but the vagina is capable of much distention and elongation; this flexibility allows the passage of the baby during childbirth. The vagina is located in the lower pelvis, between the bladder and the rectum, and is supported by the cervical ligaments and pelvic floor muscles.

THE VULVA

The female external genitalia are collectively known as the vulva. They consist of paired folds, the labia majora, which are covered by skin and pubic hair and have a moist internal lining. The labia minora are fleshy folds which lie on either side of the vestibule containing the vaginal and urinary openings and mucus-secreting glands. At the junction of the vagina and the labia minora is a thin fibrous membrane, called the hymen; the upper ends of the labia minora join around the clitoris. On either side of the vaginal opening, at the innermost part of the labia, are the Bartholin's glands.

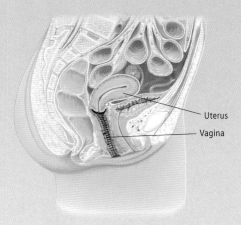

Uterus

Vagina

Vagina

The vagina is a muscular tube that extends from the cervix to the vulva, connecting the uterus to the outside of the body. Situated in the lower pelvis between the bladder and rectum, the walls of the vagina are supported by the cervical ligaments and pelvic floor muscles.

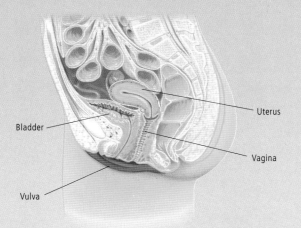

Uterus

Bladder

Vagina

Vulva

Vulva

The vulva is the collective term for the female external genitalia, which includes the labia majora, the labia minora and the clitoris.

The Shoulders, Arms and Hands

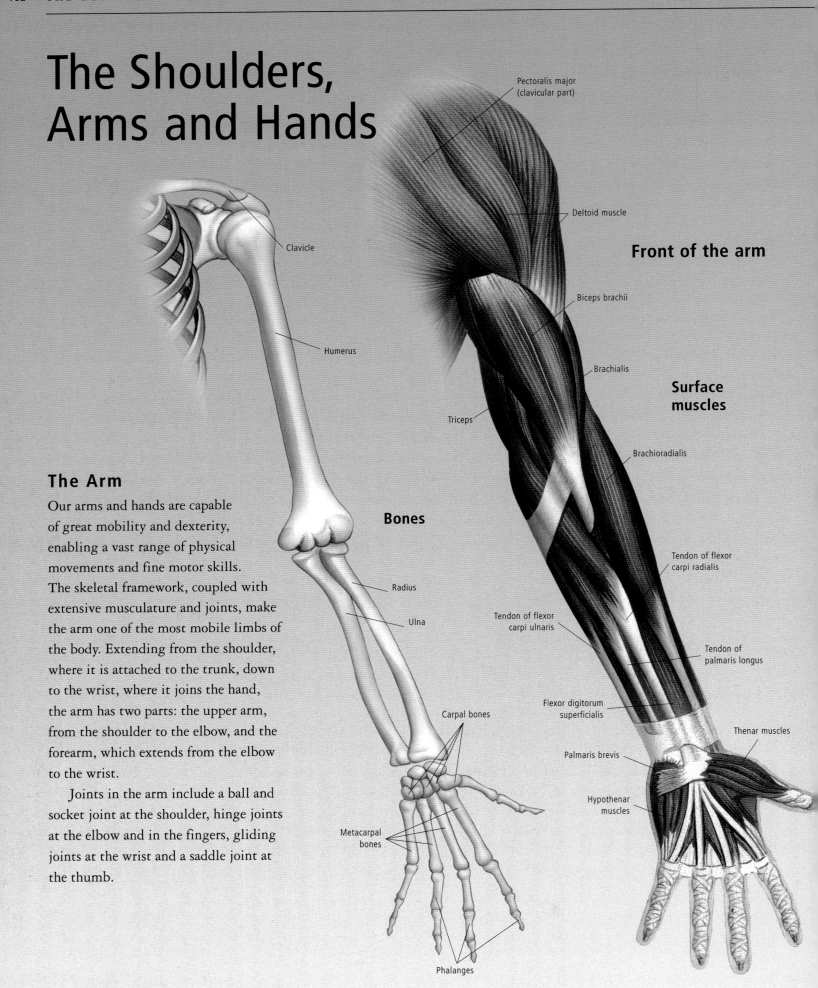

Pectoralis major (clavicular part)

Clavicle

Humerus

Front of the arm

Deltoid muscle

Biceps brachii

Brachialis

Surface muscles

Triceps

Brachioradialis

Bones

Radius

Ulna

Tendon of flexor carpi radialis

Tendon of flexor carpi ulnaris

Tendon of palmaris longus

Flexor digitorum superficialis

Carpal bones

Thenar muscles

Palmaris brevis

Hypothenar muscles

Metacarpal bones

Phalanges

The Arm

Our arms and hands are capable of great mobility and dexterity, enabling a vast range of physical movements and fine motor skills. The skeletal framework, coupled with extensive musculature and joints, make the arm one of the most mobile limbs of the body. Extending from the shoulder, where it is attached to the trunk, down to the wrist, where it joins the hand, the arm has two parts: the upper arm, from the shoulder to the elbow, and the forearm, which extends from the elbow to the wrist.

Joints in the arm include a ball and socket joint at the shoulder, hinge joints at the elbow and in the fingers, gliding joints at the wrist and a saddle joint at the thumb.

Back of the arm

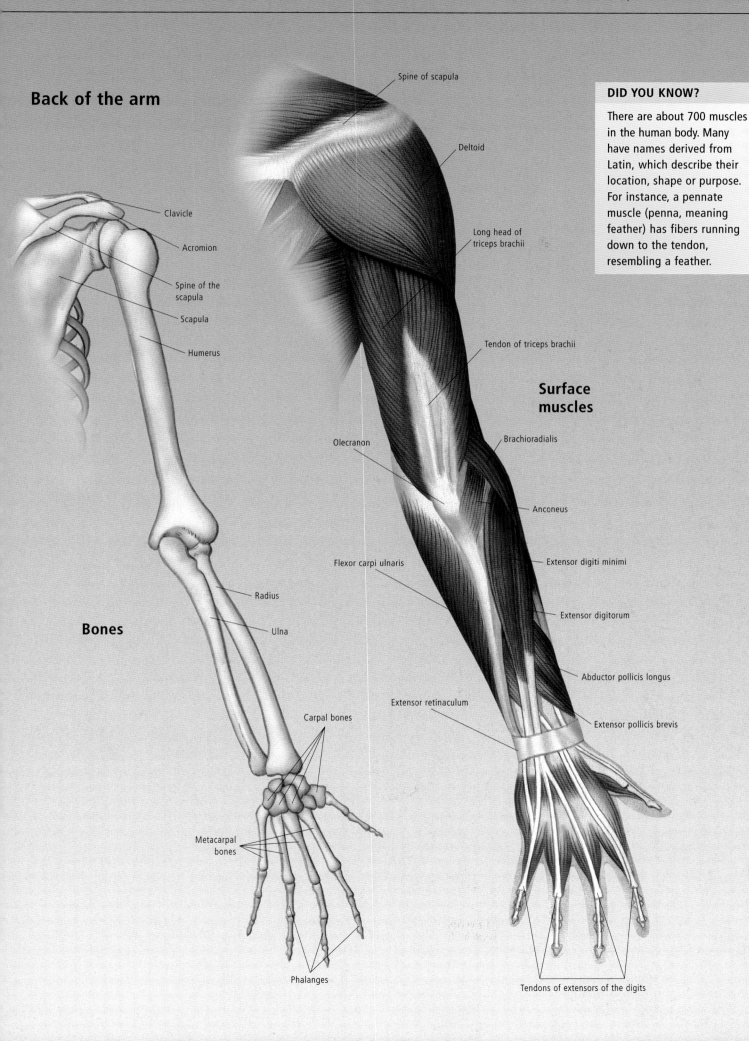

DID YOU KNOW?

There are about 700 muscles in the human body. Many have names derived from Latin, which describe their location, shape or purpose. For instance, a pennate muscle (penna, meaning feather) has fibers running down to the tendon, resembling a feather.

Spine of scapula

Deltoid

Clavicle

Acromion

Long head of triceps brachii

Spine of the scapula

Scapula

Humerus

Tendon of triceps brachii

Surface muscles

Brachioradialis

Olecranon

Anconeus

Extensor digiti minimi

Flexor carpi ulnaris

Extensor digitorum

Bones

Radius

Ulna

Abductor pollicis longus

Extensor retinaculum

Extensor pollicis brevis

Carpal bones

Metacarpal bones

Phalanges

Tendons of extensors of the digits

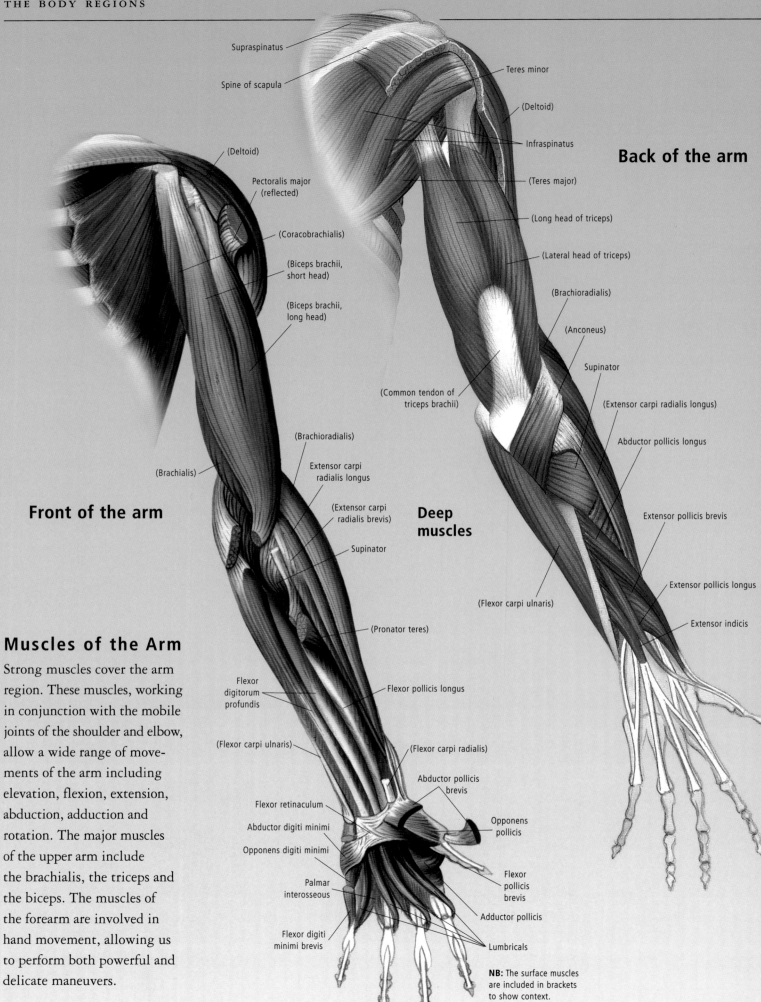

Supraspinatus

Spine of scapula

Teres minor

(Deltoid)

Back of the arm

Infraspinatus

(Deltoid)

Pectoralis major
(reflected)

(Teres major)

(Long head of triceps)

(Lateral head of triceps)

(Coracobrachialis)

(Biceps brachii,
short head)

(Brachioradialis)

(Anconeus)

(Biceps brachii,
long head)

Supinator

(Extensor carpi radialis longus)

Abductor pollicis longus

(Common tendon of
triceps brachii)

(Brachioradialis)

Front of the arm

Extensor carpi
radialis longus

(Brachialis)

(Extensor carpi
radialis brevis)

**Deep
muscles**

Extensor pollicis brevis

Supinator

Extensor pollicis longus

(Flexor carpi ulnaris)

Extensor indicis

(Pronator teres)

Muscles of the Arm

Strong muscles cover the arm
region. These muscles, working
in conjunction with the mobile
joints of the shoulder and elbow,
allow a wide range of movements of the arm including
elevation, flexion, extension,
abduction, adduction and
rotation. The major muscles
of the upper arm include
the brachialis, the triceps and
the biceps. The muscles of
the forearm are involved in
hand movement, allowing us
to perform both powerful and
delicate maneuvers.

Flexor
digitorum
profundis

Flexor pollicis longus

(Flexor carpi ulnaris)

(Flexor carpi radialis)

Abductor pollicis
brevis

Flexor retinaculum

Opponens
pollicis

Abductor digiti minimi

Opponens digiti minimi

Flexor
pollicis
brevis

Palmar
interosseous

Adductor pollicis

Flexor digiti
minimi brevis

Lumbricals

NB: The surface muscles
are included in brackets
to show context.

Blood Vessels and Nerves of the Arm

Blood vessels course through the arm, the major artery being the brachial artery. This branches at the elbow into the radial and ulnar arteries of the forearm. Either of these arteries can be monitored when the pulse is taken. These two arteries then branch into the digital arteries that supply the fingers.

Major veins draining blood from the arm include the subclavian vein, found in the neck area, and the axillary vein (found in the axilla, or armpit).

The nerves of the arm lead from the brachial plexus in the neck, down through the axilla (armpit) and into the arm. Important nerves of the arm include the axillary nerve, the radial nerve and the ulnar nerve.

Ulnar nerve

Running the length of the arm, the ulnar nerve supplies the flexor muscles of the forearm, small muscles of the hand and skin on the little finger side of the hand.

Ulnar nerve

Radial nerve

The radial nerve extends down the arm, supplying the extensor muscles in the back of the arm, the wrist and hand, and the skin of the arm and hand, on the thumb side.

Radial nerve

Radial artery

Radial artery

The radial artery supplies blood to the muscles of the forearm. It runs close to the surface of the skin at the wrist, and is often used as a pulse point.

Subclavian vein

Subclavian vein

The subclavian vein, one of the major veins in the upper body, lies beneath the collar bone. It drains blood from the arm and carries it towards the heart.

Bones of the Shoulder and Arm

The arm bones are the humerus, the radius and the ulna. The humerus is connected to the shoulder blade (scapula) at the shoulder joint. The two bones of the forearm—the radius and ulna—extend from the elbow to the wrist.

Humerus

The humerus is a long bone, cylindrical in shape, with two enlarged extremities. At its upper extremity it is joined at the shoulder joint to the scapula (shoulder blade). At its lower end, it articulates with the radius and ulna to form the elbow joint.

Greater tubercle

Anatomical neck

Head of humerus

Lesser tubercle

Surgical neck

Humerus

Acromion

Coracoid process

Suprascapular notch

Glenoid fossa

Scapula

A triangular, flat plate of bone, the shoulder blade (scapula) has several bony projections and a raised spine. The projections at its outer end join with the rounded head of the humerus to form a mobile ball-and-socket joint. Also at the outer end is the acromioclavicular joint, which joins the scapula to the collar bone (clavicle).

Subscapular fossa

Scapula

Clavicle

Spine of scapula

Acromion

Glenoid fossa

Humerus

Scapula—back view

Each scapula is connected to the humerus and clavicle (collar bone).

Medial condyle

Medial epicondyle

Lateral condyle

Lateral epicondyle

Trochlea

Head of radius

Olecranon

Trochlear notch

Coronoid process

Radius

The radius works in conjunction with the ulna, to move the forearm. The rounded head of the radius forms a pivot joint at the elbow, which allows it to rotate around the ulna and move the wrist. The radius lies on the thumb side of the forearm.

Radial tuberosity

Radial notch of ulna

Ulnar tubercle

Ulnar tuberosity

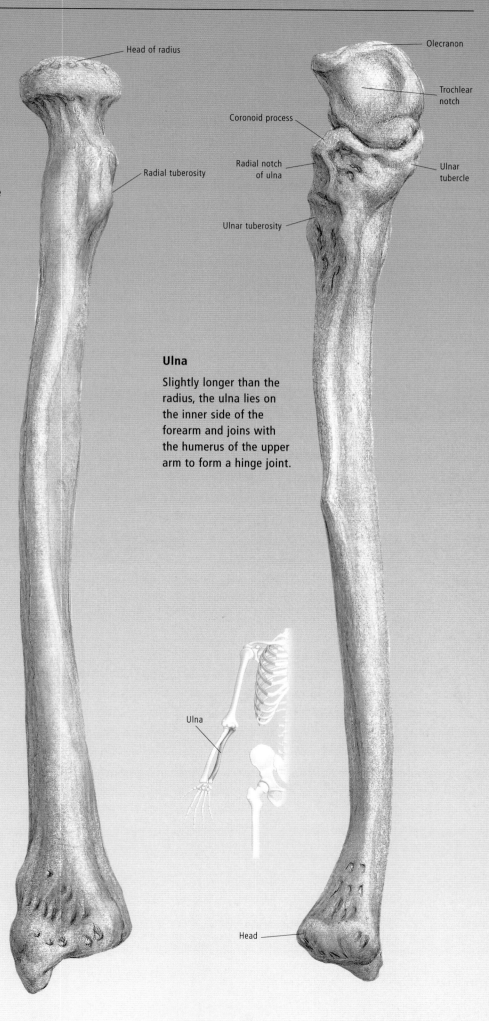

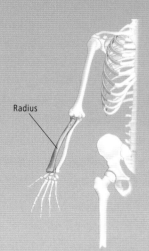

Radius

Ulna

Slightly longer than the radius, the ulna lies on the inner side of the forearm and joins with the humerus of the upper arm to form a hinge joint.

Ulna

THE UPPER ARM

The upper arm contains only one bone, the humerus. It is supported by powerful muscles such the triceps, biceps and deltoid muscles, which help to elevate and rotate the arm.

THE FOREARM

There are two bones in the forearm, the radius and the ulna. These bones extend between the elbow joint and the wrist. The muscles of the forearm are involved in many of the movements of the hand.

Head

The Shoulder

The shoulder, with its powerful muscles and versatile joint, provides the arm with a wide range of movements. The interconnected bones of the collar bone (clavicle), the shoulder blade (scapula) and humerus form the shoulder. The shoulder blade connects to the humerus of the arm at the shoulder joint, a ball-and-socket joint, and to the collar bone at the acromioclavicular joint, a gliding joint. The ball-and-socket joint of the shoulder, the most mobile joint in the body, is surrounded by a membrane capsule filled with synovial fluid. This capsule cushions the joint area and allows for smooth movements. Strong muscles power the shoulder area; these assist with the many movements required, and provide stability to this highly mobile area. The shoulder muscles are divided into two groups: those attaching the humerus to the pectoral girdle and those attaching the pectoral girdle to the trunk.

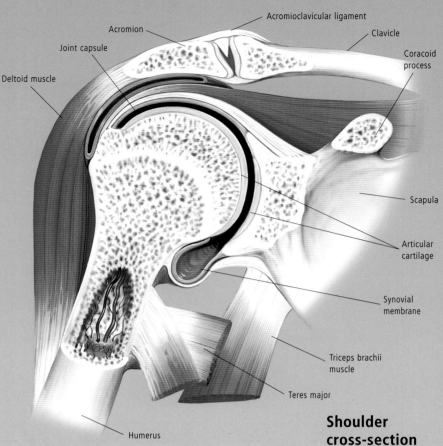

Acromioclavicular ligament

Acromion

Clavicle

Joint capsule

Coracoid process

Deltoid muscle

Scapula

Articular cartilage

Synovial membrane

Triceps brachii muscle

Teres major

Humerus

Shoulder cross-section

Rotator cuff muscles

These powerful muscles stabilize the shoulder, holding the head of the humerus firmly in place in the shoulder socket. Comprised of the subscapularis at the front and the supraspinatus, infraspinatus and teres minor at the back, these muscles arise from the scapula and attach to the humerus.

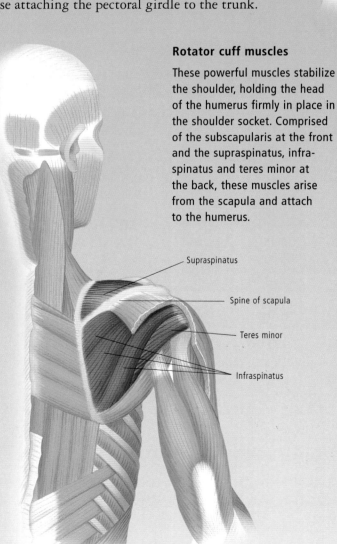

Supraspinatus

Spine of scapula

Teres minor

Infraspinatus

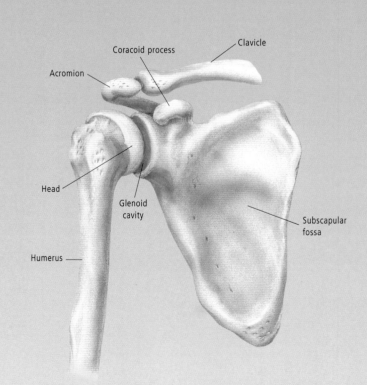

Coracoid process

Clavicle

Acromion

Head

Glenoid cavity

Subscapular fossa

Humerus

Shoulder joint

The great flexibility of the shoulder is due to its highly mobile ball-and-socket joint. The head of the humerus (ball) sits in the glenoid cavity of the scapula (socket). Shoulder movements always occur in conjunction with movements of the pectoral girdle (clavicle and scapula).

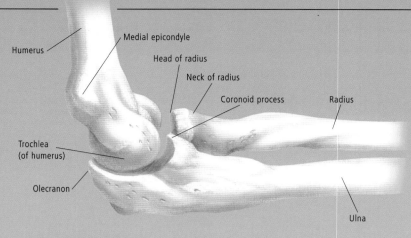

Humerus

Medial epicondyle

Head of radius

Neck of radius

Coronoid process

Radius

Trochlea (of humerus)

Olecranon

Ulna

The Elbow

Connecting the humerus of the upper arm to the radius and ulna of the forearm is the hinge joint of the elbow. Bony protrusions at the end of the humerus, known as epicondyles, articulate with the radius and ulna. Ligaments keep the area stable, while muscles in the upper arm move the elbow: the biceps muscle bending the elbow, and the triceps muscle straightening it out. The relationship between the radius and ulna also permits some rotational movement of the elbow, with the radius rotating around the ulna, thus turning the hands.

Elbow Movement

The muscles required to make a joint move are arranged in pairs. One muscle allows flexion and its partner allows extension. Elbow movements are controlled by the biceps and triceps muscles; the biceps controls flexion of the elbow, while the triceps controls elbow extension.

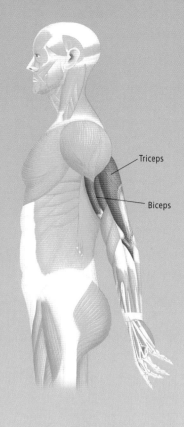

Triceps

Biceps

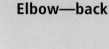

Elbow—back

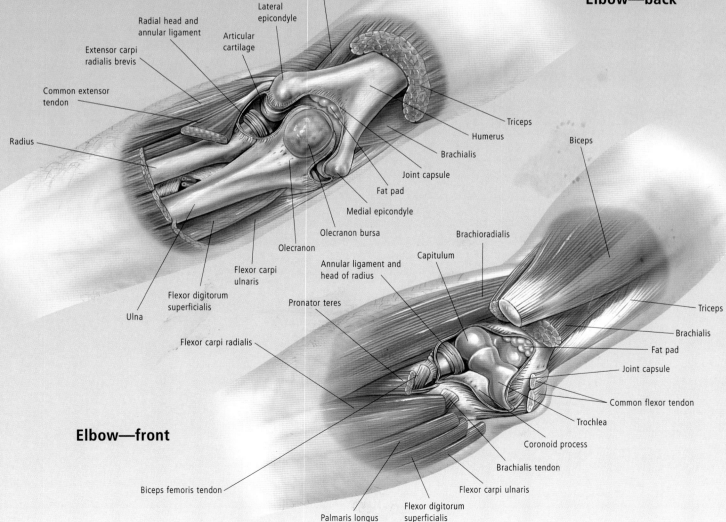

Brachioradialis

Lateral epicondyle

Radial head and annular ligament

Articular cartilage

Extensor carpi radialis brevis

Common extensor tendon

Radius

Triceps

Humerus

Brachialis

Joint capsule

Fat pad

Medial epicondyle

Olecranon bursa

Olecranon

Annular ligament and head of radius

Flexor carpi ulnaris

Flexor digitorum superficialis

Ulna

Flexor carpi radialis

Pronator teres

Biceps

Brachioradialis

Capitulum

Triceps

Brachialis

Fat pad

Joint capsule

Common flexor tendon

Trochlea

Coronoid process

Brachialis tendon

Flexor carpi ulnaris

Flexor digitorum superficialis

Palmaris longus

Biceps femoris tendon

Elbow—front

The Hand

The hand is designed to grasp and manipulate objects. It consists of the palm, or front of the hand, the dorsum, or back, and the thumb and fingers. Thick skin covers the palm, and the slightly cupped surface is designed to assist in hand movements such as gripping. The tendons in the palm flex the fingers, bending them forward, while the tendons at the back of the hand extend the fingers, pulling them backwards.

Nerves traveling down from the forearm supply the muscles and skin of the hand; the major nerves of the hand are the ulnar, median and radial nerves.

Bones of the Hand

Five metacarpal bones lead from the wrist and join to the bones of the fingers, known as the phalanges: Each of the fingers has three phalanges (proximal, middle and distal); the thumb has just two, the proximal and distal phalanges.

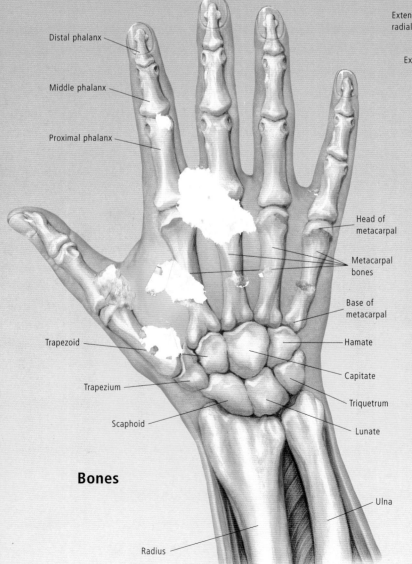

Distal phalanx

Middle phalanx

Proximal phalanx

Trapezoid

Trapezium

Scaphoid

Bones

Head of metacarpal

Metacarpal bones

Base of metacarpal

Hamate

Capitate

Triquetrum

Lunate

Ulna

Radius

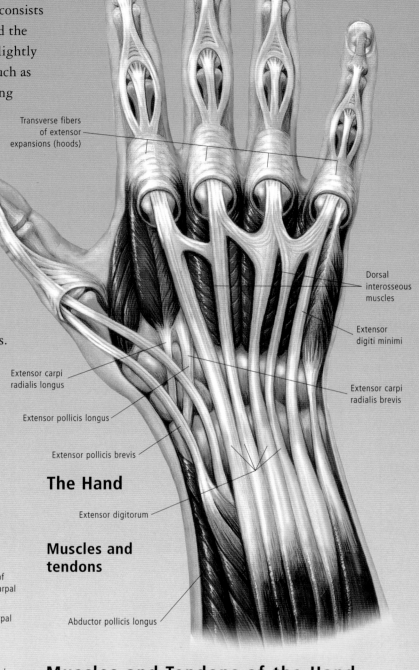

Transverse fibers of extensor expansions (hoods)

Dorsal interosseous muscles

Extensor digiti minimi

Extensor carpi radialis longus

Extensor carpi radialis brevis

Extensor pollicis longus

Extensor pollicis brevis

The Hand

Extensor digitorum

Muscles and tendons

Abductor pollicis longus

Muscles and Tendons of the Hand

Many of the movements of the hand require the activation of not only the hand muscles, but also the powerful muscles of the arm. Depending on the action required, either the power grip used to pick up an object, or the precision grip used in holding a pen, different muscles spring into action. The thenar muscles create the soft bulge of the ball of the thumb; these muscles and a separate adductor muscle contribute to the movement of the thumb. The hypothenar muscles create the fleshy bulge at the side of the palm. Between each of the metacarpal bones lie the interosseous muscles, associated with many movements of the fingers.

The Wrist

The wrist joint, where the radius and ulna join with the carpal bones of the wrist, allows a wide range of movements to be performed, including flexion, extension and sideways movement. Passing over and under the carpal bones of the wrist are the tendinous sheaths of connective tissue through which the tendons to the fingers and thumb pass. Nerves and blood vessels pass over and supply the wrist area. The ulnar and radial arteries of the forearm cross the wrist before branching into the digital arteries of the fingers. The ulnar, median and radial nerves also pass over the wrist area.

The Fingers

As there are no muscles in the fingers, the dexterity of the fingers occurs as a combined operation between the muscles of the arm and hand, and the tendons in the fingers. Each of the four fingers has three bones or phalanges: the proximal, middle and distal phalanges. The phalanges are joined to one another by a hinge joint, and the proximal phalanges, those closest to the palm, are then connected by another hinge joint to the metacarpal bones of the hand. The fingertips are protected on the outer (dorsal) side by the fingernails, while on the palm side the fingertips are covered in various patterns of ridges and whorls that give us the unique finger-prints that identify each person, since no two are alike.

DID YOU KNOW?

From delicate precision tasks to powerful gripping actions, our hands are constantly busy, and during the average lifetime the finger joints will flex over 25 million times.

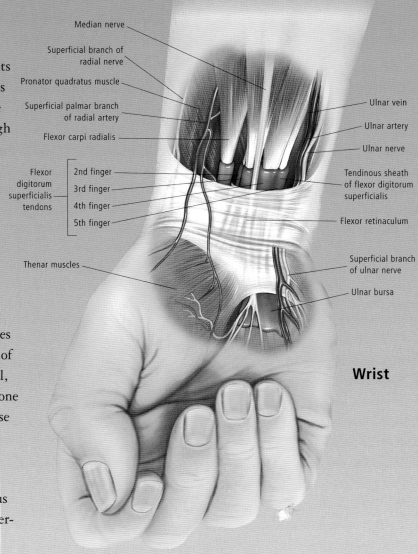

Median nerve
Superficial branch of radial nerve
Pronator quadratus muscle
Superficial palmar branch of radial artery
Flexor carpi radialis
Flexor digitorum superficialis tendons
  2nd finger
  3rd finger
  4th finger
  5th finger
Thenar muscles
Ulnar vein
Ulnar artery
Ulnar nerve
Tendinous sheath of flexor digitorum superficialis
Flexor retinaculum
Superficial branch of ulnar nerve
Ulnar bursa

Wrist

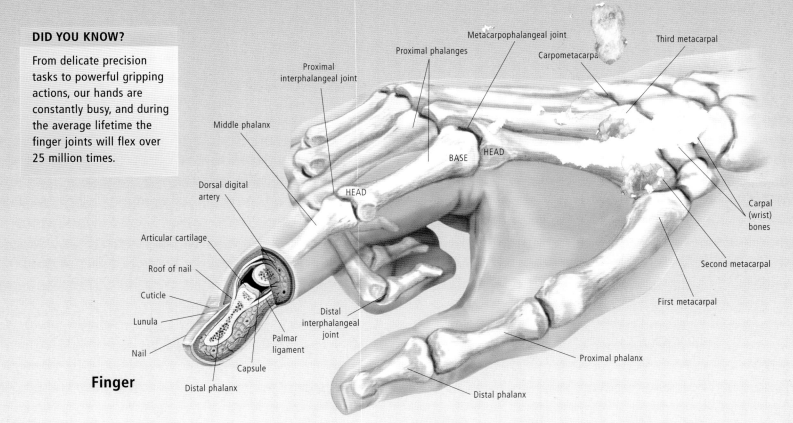

Metacarpophalangeal joint
Proximal phalanges
Third metacarpal
Carpometacarpal
Proximal interphalangeal joint
Middle phalanx
BASE   HEAD
HEAD
Dorsal digital artery
Articular cartilage
Roof of nail
Cuticle
Lunula
Nail
Palmar ligament
Distal interphalangeal joint
Capsule
Carpal (wrist) bones
Second metacarpal
First metacarpal
Proximal phalanx
Distal phalanx

Finger

Distal phalanx

The Hips, Legs and Feet

The front of the leg

The Leg

Connected to the hip bones at the top, and the foot at the bottom, the three long bones of the leg are the femur, tibia and fibula. Although in anatomical terms, the leg refers only to the area between the knee and the ankle, the popular usage of the term "leg" refers to the whole limb, except for the foot. The hip bone forms the connection between the spine and the lower limbs. The lower limbs must carry the weight of the torso, therefore large powerful muscles provide stability and generate the leg movements. Large joints at the hip and knee and strong ligaments throughout the leg complete the structural mechanism for bipedal locomotion, that is, walking upright on two legs.

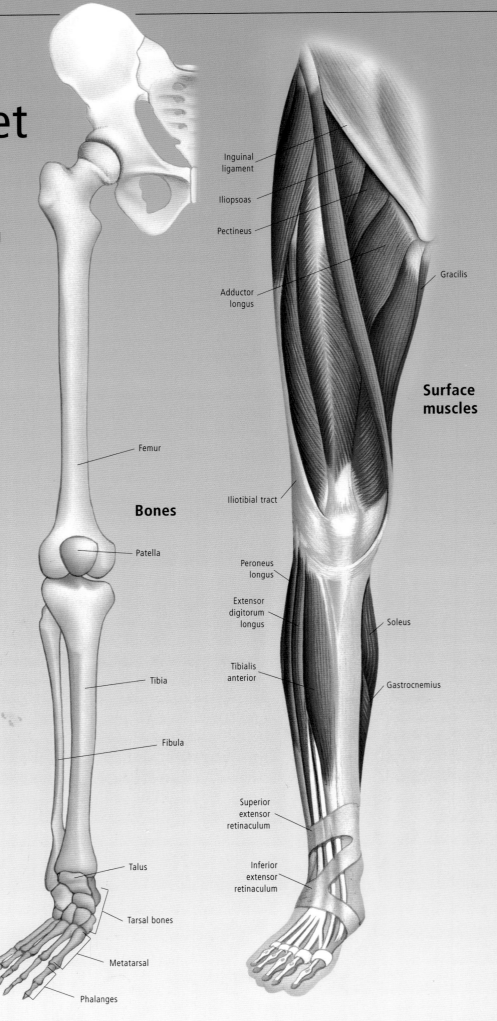

Bones

Femur

Patella

Tibia

Fibula

Talus

Tarsal bones

Metatarsal

Phalanges

Surface muscles

Inguinal ligament

Iliopsoas

Pectineus

Adductor longus

Gracilis

Iliotibial tract

Peroneus longus

Extensor digitorum longus

Soleus

Tibialis anterior

Gastrocnemius

Superior extensor retinaculum

Inferior extensor retinaculum

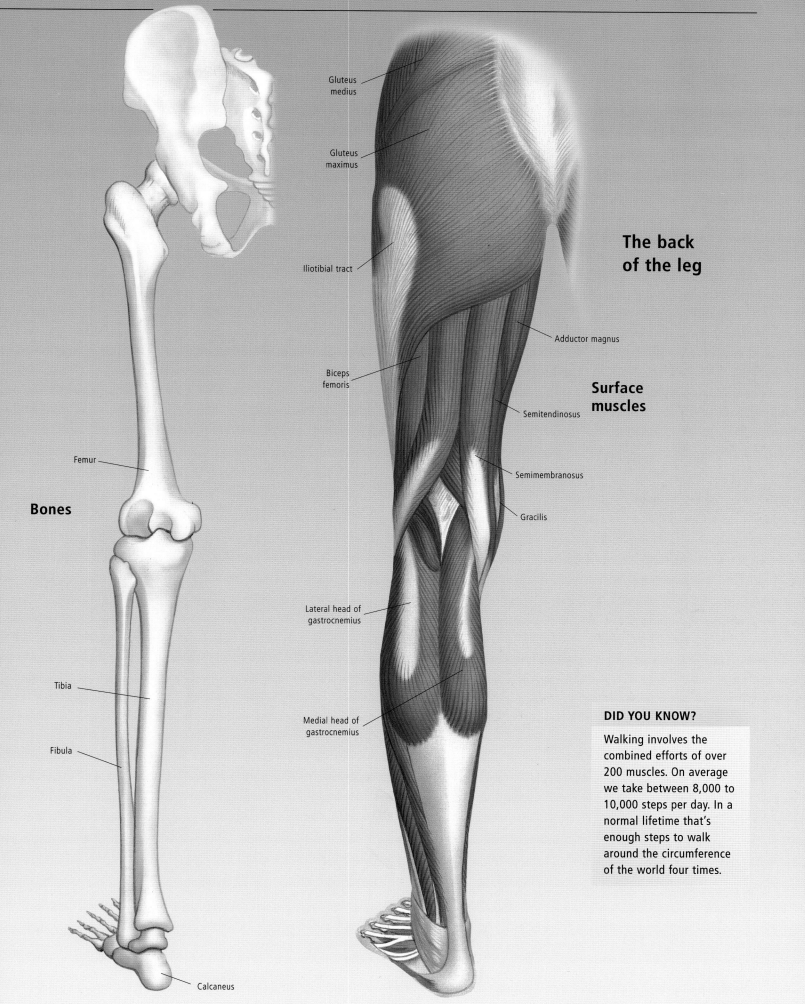

Gluteus medius

Gluteus maximus

Iliotibial tract

The back of the leg

Adductor magnus

Biceps femoris

Surface muscles

Semitendinosus

Semimembranosus

Gracilis

Bones

Femur

Lateral head of gastrocnemius

Tibia

Medial head of gastrocnemius

Fibula

DID YOU KNOW?

Walking involves the combined efforts of over 200 muscles. On average we take between 8,000 to 10,000 steps per day. In a normal lifetime that's enough steps to walk around the circumference of the world four times.

Calcaneus

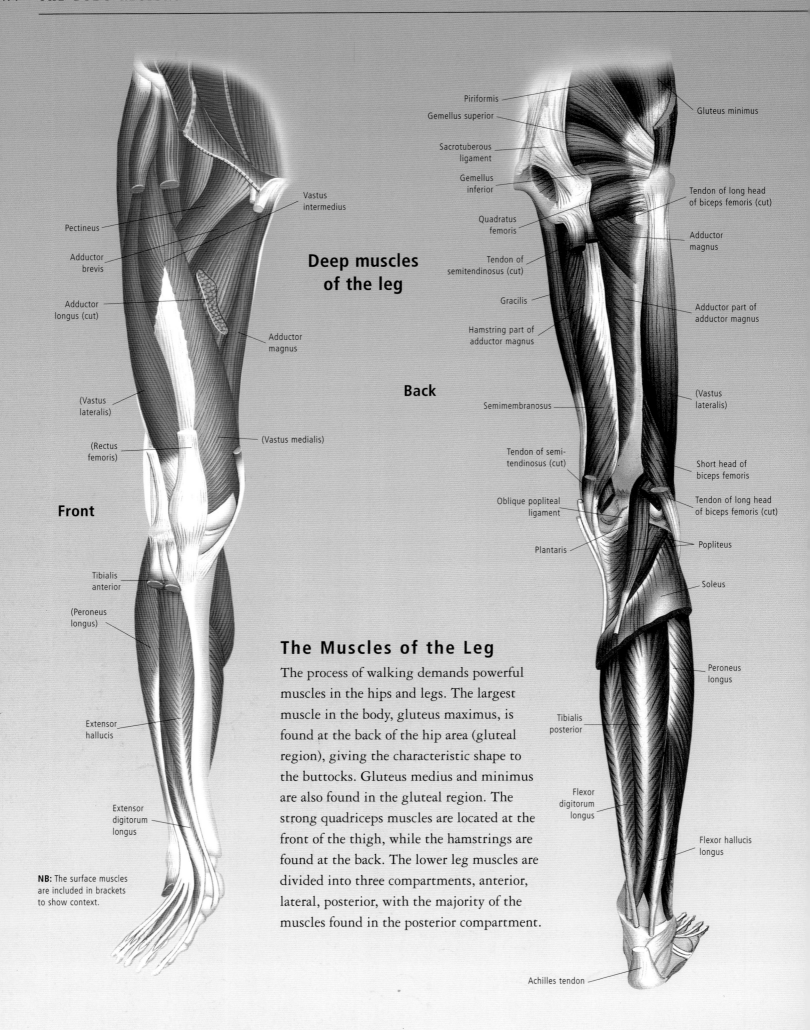

Pectineus

Adductor
brevis

Adductor
longus (cut)

Vastus
intermedius

**Deep muscles
of the leg**

Adductor
magnus

(Vastus
lateralis)

(Rectus
femoris)

(Vastus medialis)

Front

Tibialis
anterior

(Peroneus
longus)

Extensor
hallucis

Extensor
digitorum
longus

NB: The surface muscles
are included in brackets
to show context.

Piriformis

Gemellus superior

Sacrotuberous
ligament

Gemellus
inferior

Quadratus
femoris

Tendon of
semitendinosus (cut)

Gracilis

Hamstring part of
adductor magnus

Back

Semimembranosus

Tendon of semi-
tendinosus (cut)

Oblique popliteal
ligament

Plantaris

Gluteus minimus

Tendon of long head
of biceps femoris (cut)

Adductor
magnus

Adductor part of
adductor magnus

(Vastus
lateralis)

Short head of
biceps femoris

Tendon of long head
of biceps femoris (cut)

Popliteus

Soleus

Peroneus
longus

Tibialis
posterior

Flexor
digitorum
longus

Flexor hallucis
longus

Achilles tendon

The Muscles of the Leg

The process of walking demands powerful
muscles in the hips and legs. The largest
muscle in the body, gluteus maximus, is
found at the back of the hip area (gluteal
region), giving the characteristic shape to
the buttocks. Gluteus medius and minimus
are also found in the gluteal region. The
strong quadriceps muscles are located at the
front of the thigh, while the hamstrings are
found at the back. The lower leg muscles are
divided into three compartments, anterior,
lateral, posterior, with the majority of the
muscles found in the posterior compartment.

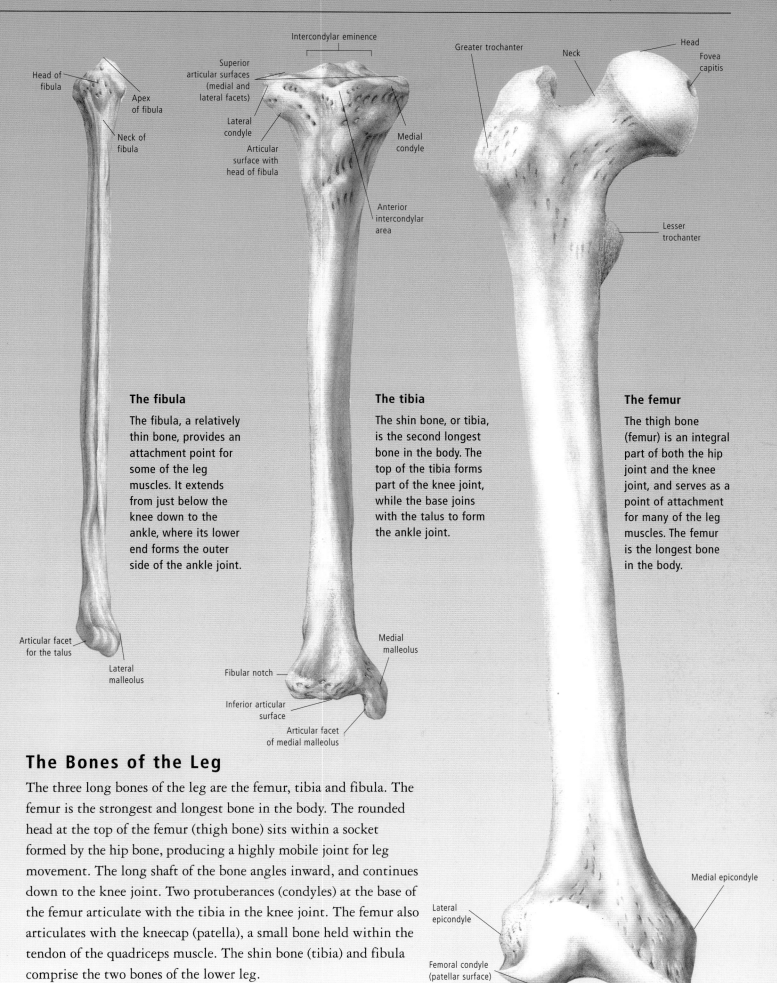

Intercondylar eminence

Head of fibula

Apex of fibula

Neck of fibula

Superior articular surfaces (medial and lateral facets)

Lateral condyle

Articular surface with head of fibula

Medial condyle

Anterior intercondylar area

Greater trochanter

Neck

Head

Fovea capitis

Lesser trochanter

Articular facet for the talus

Lateral malleolus

Medial malleolus

Fibular notch

Inferior articular surface

Articular facet of medial malleolus

Medial epicondyle

Lateral epicondyle

Femoral condyle (patellar surface)

The fibula

The fibula, a relatively thin bone, provides an attachment point for some of the leg muscles. It extends from just below the knee down to the ankle, where its lower end forms the outer side of the ankle joint.

The tibia

The shin bone, or tibia, is the second longest bone in the body. The top of the tibia forms part of the knee joint, while the base joins with the talus to form the ankle joint.

The femur

The thigh bone (femur) is an integral part of both the hip joint and the knee joint, and serves as a point of attachment for many of the leg muscles. The femur is the longest bone in the body.

The Bones of the Leg

The three long bones of the leg are the femur, tibia and fibula. The femur is the strongest and longest bone in the body. The rounded head at the top of the femur (thigh bone) sits within a socket formed by the hip bone, producing a highly mobile joint for leg movement. The long shaft of the bone angles inward, and continues down to the knee joint. Two protuberances (condyles) at the base of the femur articulate with the tibia in the knee joint. The femur also articulates with the kneecap (patella), a small bone held within the tendon of the quadriceps muscle. The shin bone (tibia) and fibula comprise the two bones of the lower leg.

Specialized Muscles of the Leg

CALF MUSCLES

The calf region contains compartments of muscles responsible for various movements of the foot. The muscles of the anterior compartment move the foot upwards; the muscles of the lateral compartment turn out the sole of the foot; the superficial muscles of the posterior compartment move the foot downwards, while the deeper muscles that pass behind the ankle joint are crucial to pushing off from the big toe.

QUADRICEPS MUSCLES

Quadriceps femoris is a muscle formed by four separate parts, namely the rectus femoris, vastus lateralis, vastus medialis and vastus intermedius. The upper end of the quadriceps muscle arises in the pelvis and upper thigh bone. At its lower end, the tendons of the four components merge together, and attach to the kneecap (patella), then extend down to their point of attachment on the tibia. The quadriceps flexes the thigh, straightens the knee, and provides strength and stability to the knee joint.

HAMSTRING MUSCLES

Biceps femoris, semitendinosus and semimembranous comprise the hamstring muscles, which mass at the back of the thigh. These surface muscles extend from their point of attachment in the lower leg, passing to the back of the knee and attach at the top to the ischium at the base of the pelvis. This group of strong muscles is involved in the movement of both the hip and knee joints.

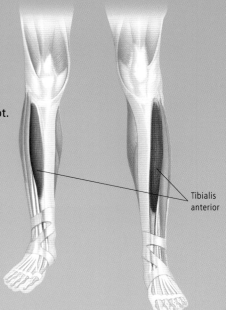

Calf muscles
The tibialis anterior muscle attaches to the tibia and allows inward and upward movements of the foot.

Tibialis anterior

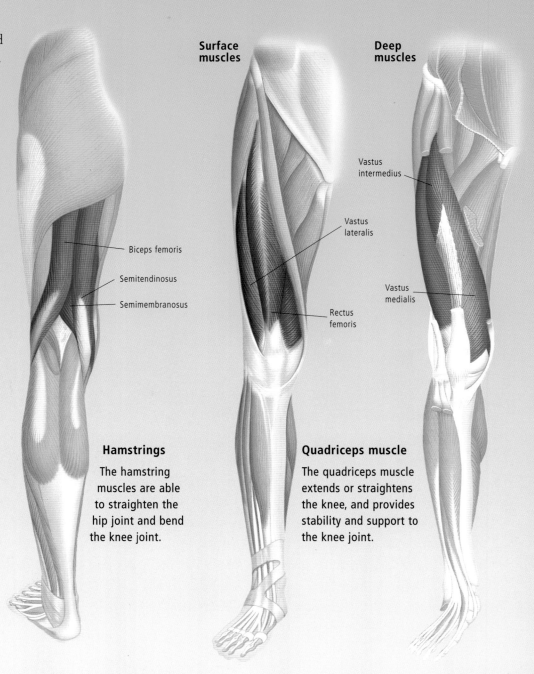

Surface muscles

Deep muscles

Biceps femoris

Semitendinosus

Semimembranosus

Vastus intermedius

Vastus lateralis

Vastus medialis

Rectus femoris

Hamstrings
The hamstring muscles are able to straighten the hip joint and bend the knee joint.

Quadriceps muscle
The quadriceps muscle extends or straightens the knee, and provides stability and support to the knee joint.

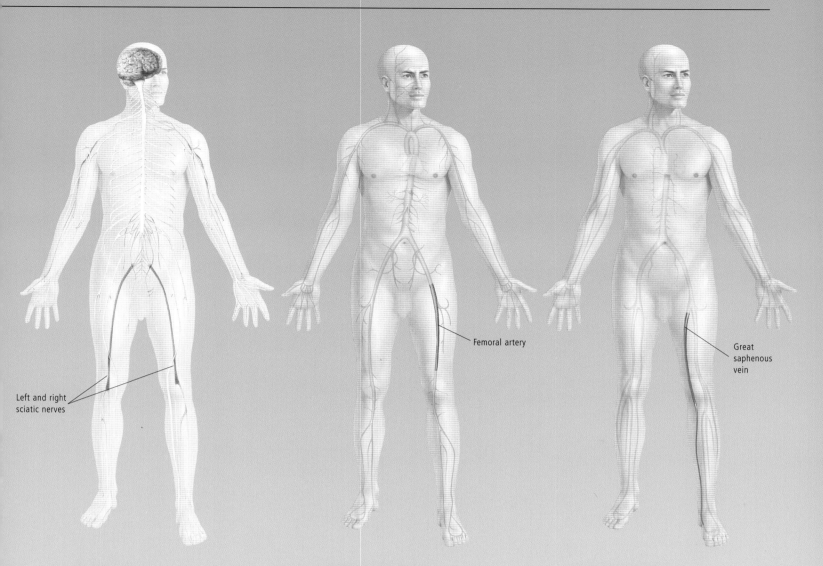

Left and right
sciatic nerves

Femoral artery

Great
saphenous
vein

Sciatic nerve

The thickest nerve of the body, the sciatic nerve extends from the base of the spine down the thigh, then branches out through the lower leg and foot. It supplies many muscles, including the hamstrings.

Femoral artery

The external iliac artery in the trunk becomes the femoral artery as it passes through the groin area, where it can be felt as a pulse point. The femoral artery supplies the legs and hips.

Saphenous vein

Extending from the foot to the groin, the great saphenous vein is the longest vein in the body. Blood from the vein drains into the deep veins of the leg, which return blood to the heart.

Nerves and Blood Vessels of the Leg

Large arteries and veins supply a network of smaller blood vessels in the leg. The main artery supplying the leg is the femoral artery. This artery, supplied by the aorta via the external iliac artery, runs part-way down the thigh before running behind the knee, where it becomes the popliteal artery. This then branches into the various compartments of the lower leg.

Superficial and deep veins run through the legs. The deep veins follow the arteries, while the superficial veins run just below the skin. An extensive valve system operates in the veins of the leg. As the leg muscles expand and contract, they squeeze and relax the valves. When the valves are squeezed, blood is forced upwards towards the heart. When the muscles relax, blood is allowed to flow from the superficial veins to the deep veins. The great saphenous vein, the principal superficial vein of the leg, is the longest vein in the body.

The major nerves supplying the leg are the femoral, obturator and sciatic nerves. These nerves, subdividing into smaller nerves, supply the muscles and skin of the legs. The sciatic nerve supplies the hamstring muscles, lower leg muscles and the foot.

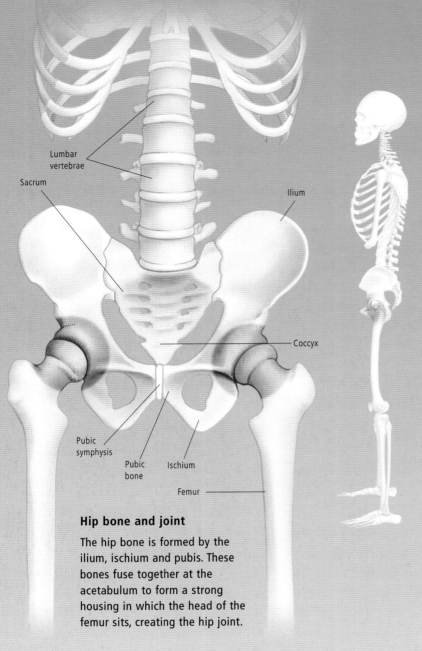

Lumbar
vertebrae

Sacrum

Ilium

Coccyx

Pubic
symphysis

Pubic
bone

Ischium

Femur

Hip bone and joint

The hip bone is formed by the
ilium, ischium and pubis. These
bones fuse together at the
acetabulum to form a strong
housing in which the head of the
femur sits, creating the hip joint.

The Hip Bone and Hip Joint

Joined to the spine at the sacroiliac joint, the hip
consists of the ilium, ischium and pubis. During the
early teenage years, these bones fuse together at the
acetabulum to consolidate the hip bone. The combination of the hip bone and the sacrum creates a
circle of bone around the base of the abdominal organs.

The cupped shape of the acetabulum provides the
housing for the smooth head of the femur, together
forming the hip joint. More than half the head of the
femur sits in the acetabulum, creating a stable joint
with great load-bearing capacity, yet with the capa-
bility of a wide range of movement. The head of the
femur has a cartilage covering and is lubricated by
synovial fluid, enabling friction-free, smooth movement of the joint. The joint is then encased in a
fibrous capsule that firmly secures the head of the
femur in place while allowing freedom of movement.
The strong muscles of the hip joint include the gluteus
maximus, gluteus medius and gluteus minimus at the
back and rectus femoris at the front. These muscles
hold the joint in place and move the thigh.

Ball-and-socket joint

The rounded head of the femur sits in the cup-
shaped socket of the acetabulum, creating a
ball-and-socket joint that allows freedom of
movement second only to the shoulder joint.

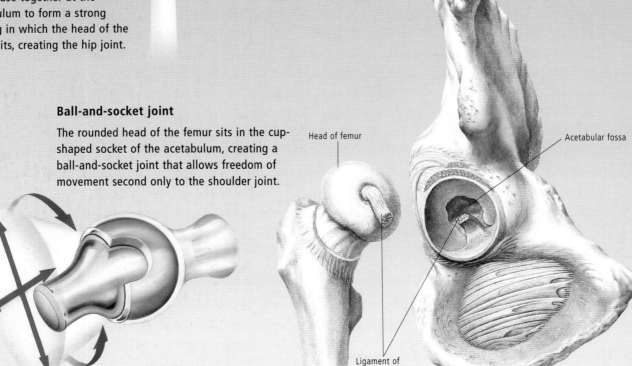

Head of femur

Acetabular fossa

Ligament of
head of femur

Knee joint

The knee is a complex hinge joint between the femur (thigh bone), the tibia (shin bone) and the patella (the kneecap). A fibrous capsule encases the joint, uniting the three bones. In the joint cavity, two wedge-shaped cartilage disks (menisci) are attached to the upper surface of the tibia, allowing a small amount of rotational movement.

Femur

Articular cartilage

Patella

Fibula

Tibia

Hinge joint

A hinge joint allows movement in one plane only, such as bending and straightening of the knee.

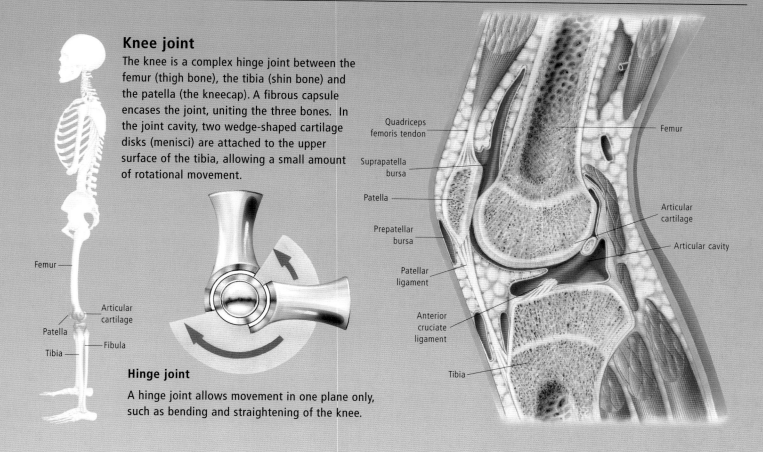

Quadriceps femoris tendon

Suprapatella bursa

Patella

Prepatellar bursa

Patellar ligament

Anterior cruciate ligament

Tibia

Femur

Articular cartilage

Articular cavity

Ligaments of the knee

The medial collateral ligament and the lateral collateral ligament reinforce the sides of the knee. Within the joint capsule are the cruciate ligaments which guide the tibia in its movement. Here, the patella has been separated to show the inner ligaments and bones of the knee joint.

Lateral condyle of femur

Fibular collateral ligament

Lateral meniscus

Posterior cruciate ligament

Anterior cruciate ligament

Medial condyle of femur

Medial meniscus

Tibial collateral ligament

Patella

The Knee and Knee Joint

One of the most complex joints in the body, the knee joins the thigh and the lower leg. Three bones form the joint—the femur (thigh bone), the tibia (shin bone), and the patella (kneecap). Essentially a hinge joint, though with some backward and forward gliding movement and a small amount of rotation, the knee is a highly mobile, weight-bearing joint.

The three bones of the knee, and the large joint cavity between them, are bound together by the knee joint capsule. This capsule has a membranous lining and is filled with synovial fluid to lubricate the connecting joint surfaces. In addition, the base of the femur and the head of the tibia have a cartilage covering to assist in the smooth movement of the knee. The patella, held in the tendon of the quadriceps muscle, glides over the front of the femur.

Since the knee joint bones do not fit closely together, a network of strong ligaments and muscles is necessary to provide strength and stability. The kneecap is held in place by strong ligaments; these ligaments allow a longitudinal sliding movement of the kneecap over the femur. Collateral ligaments run along the outer sides of the joint capsule, while within the capsule, the cruciate ligaments play a part in controlling movement of the shin bone and the kneecap. Equally important in maintaining stability are the muscles supporting the knee, including the hamstrings and the quadriceps.

The Foot and the Ankle Joint

The bones of the foot support the body and act as a lever during walking. The segmented bones allow the foot to adapt to different surfaces. Thick skin covers the sole of the foot, while the upper surface has thin skin covering tendons that run from the front of the lower leg to the tarsal bones and toes. The seven tarsal bones at the back of the foot (tarsus) include the talus, which forms part of the ankle joint, and the heel (calcaneus), which is the lower point of attachment for the Achilles tendon. The remaining tarsal bones sit together, forming gliding joints. Joining the

tarsal bones are the metatarsal bones, and joining these are the toe bones (phalanges). Each of the smaller toes has three phalanges (proximal, middle and distal). The big toe has only two phalanges (proximal and distal).

The ankle joint is formed by the meeting of the tibia and fibula of the lower leg with the talus of the foot. This synovial hinge joint, in conjunction with the muscles of the lower leg, allows movement of the foot. Protrusions at the end of the tibia and fibula (known as the medial malleolus and lateral malleolus) create the bony lumps of the ankle. These two malleoli, together with part of the tibia form a housing for the talus, creating a stable joint. Added stabilization is provided by the ligaments connecting the malleoli to the talus. Tendon sheaths run over the front of the ankle, guiding the tendons, which then exit the sheath and extend to their point of attachment on the foot bones.

The foot can move in the following ways: feet point upwards (dorsiflexion), feet point down (plantarflexion), sole faces inwards (inversion or supination) and sole faces outwards (eversion or pronation).

Foot—Bones

There are 26 bones in each foot, with the metatarsal bones and phalanges accounting for 19 of these, and the remainder forming the tarsus, or back part of the foot. The tarsus includes the calcaneus, which forms the heel, and the talus, which joins with the tibia and fibula to form the ankle joint.

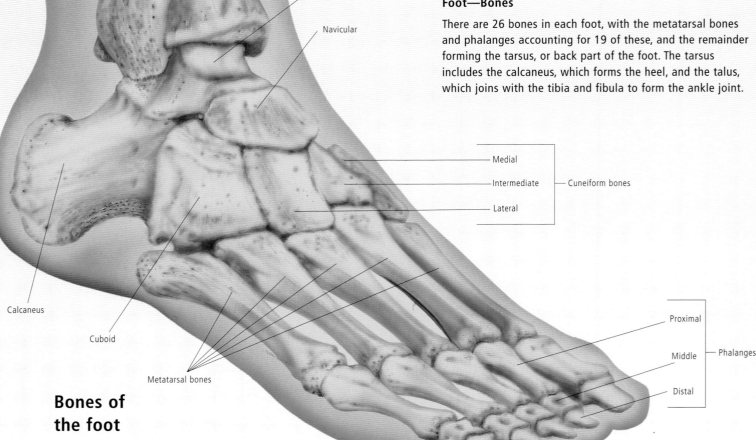

Talus

Navicular

Medial

Intermediate — Cuneiform bones

Lateral

Calcaneus

Cuboid

Proximal

Middle — Phalanges

Distal

Metatarsal bones

Bones of the foot

Foot—Muscles

The muscles of the lower leg and foot work to allow the feet to point upwards, downwards, inwards and outwards. The main muscles producing upward movement (dorsiflexion) are found at the front of the lower leg, while the muscles at the back of the leg allow downward movement (plantarflexion). Ligaments join the fibula and tibia to the tarsal bones to create a stable joint at the ankle.

Peroneus longus tendon

Peroneus brevis

Achilles (calcaneal) tendon

Superior peroneal retinaculum

Calcaneus

Inferior peroneal retinaculum

Peroneus longus

Extensor digitorum brevis

Peroneus brevis

Peroneus tertius

Extensor digitorum longus tendons

Tibialis anterior

Extensor hallucis longus

Extensor digitorum longus

Superior extensor retinaculum

Tendon sheaths

Inferior extensor retinaculum

Tendon sheath

Extensor hallucis longus

Extensor digitorum brevis tendons

Tibialis posterior muscle

Flexor digitorum longus muscle

Tibia

Flexor digitorum longus tendon

Tibialis posterior tendon

Posterior tibial artery

Tibial nerve

Flexor retinaculum

First metatarsal

Flexor hallucis longus muscle

Flexor hallucis longus tendon

Peroneus longus tendon

Achilles (calcaneal) tendon

Calcaneal tuberosity

DID YOU KNOW?

Each foot contains 26 bones, 33 joints, 107 ligaments and 19 muscles. The skin on the sole of the foot is much thicker than elsewhere on the body, and there are 125,000 sweat glands on each foot—that's more per square inch than anywhere else on the body.

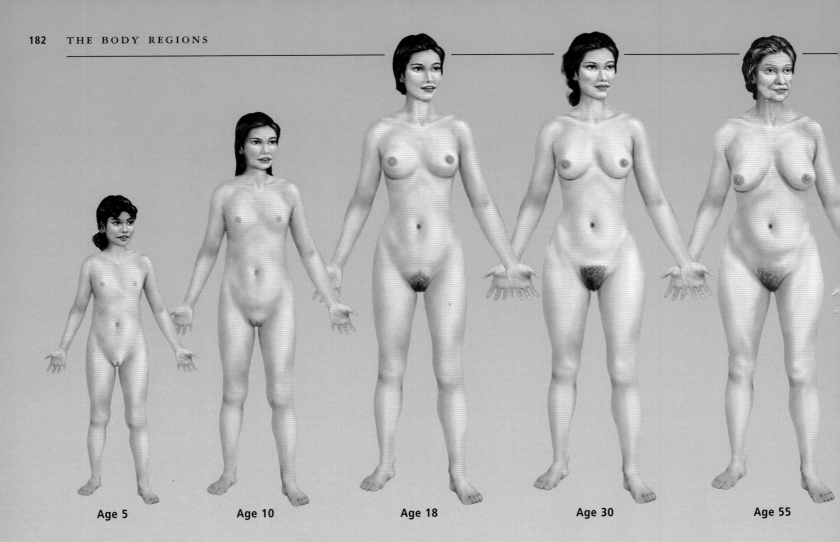

Age 5   Age 10   Age 18   Age 30   Age 55

The Human Life Cycle

Ageing

Throughout a human lifetime, many gradual yet dramatic changes take place in a person's physical appearance, and in the structure and function of the body systems.

CHILDHOOD

Between infancy and adolescence, a child goes through many developmental changes. The brain grows rapidly (it is 90 percent of its full adult size by age 6), and verbal and reasoning skills expand. The bones lengthen, and bone gradually replaces cartilage in the arms and legs. From age 6, a child begins to lose the first set of 20 baby teeth, which are then replaced by 32 permanent teeth.

ADOLESCENCE

A key feature of adolescence in both boys and girls is the growth spurt: in girls it occurs at age 10 or 11, and in boys a few years later. Height increases by 3 inches (8 centimeters) or so every year, and the body assumes more adult proportions. As adolescents mature, they are able to think more conceptually and abstractly.

   During puberty, pubic hair, underarm hair and facial hair (in boys) appear. The sweat and sebaceous glands become more active, and this can cause acne.

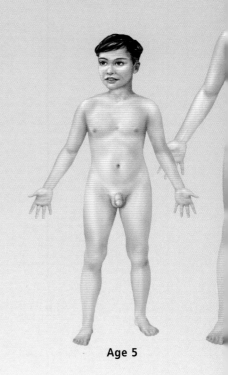

Age 5

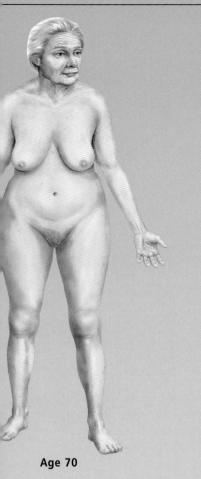

Age 70

A girl's first menstrual period occurs around 13 years of age. The ovaries increase their hormonal secretions, leading to enlargement of the breasts and external genitalia.

As boys reach puberty their muscles increase in bulk, the voice deepens and the testes grow and produce sperm. The scrotum and penis also increase in size.

ADULTHOOD AND OLD AGE

By about age 18, the human body has reached physiological, anatomical and sexual maturity, and from there it continues to change but not grow.

In women, the production of eggs and sex hormones decreases from age 35, culminating in menopause at around age 50. Without estrogen stimulation, the breasts and skin lose their firmness and elasticity, fat is distributed differently, pubic hair becomes sparse and height decreases as bones thin.

As men age, muscle development starts to decline. Hormone and sperm production is also reduced, and erections may be difficult to achieve or maintain. The skin softens, the prostate enlarges, and loss of hair is common.

In both sexes, hair turns gray due to poor functioning of pigment cells. Age spots appear on the skin, and all the systems of the body deteriorate.

A good diet, regular exercise, mental stimulation and a positive outlook may help to diminish the degenerative effects of ageing.

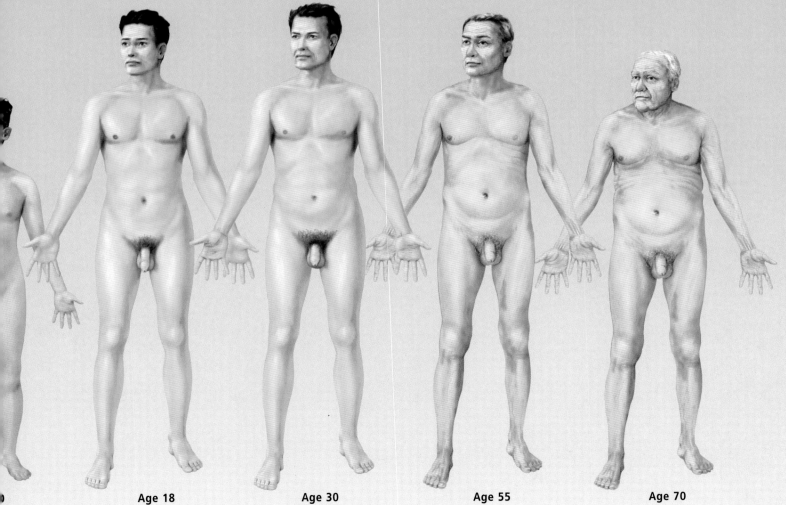

Age 18   **Age 30**   **Age 55**   **Age 70**

Fertility

Fertility, the ability to conceive a baby, is dependent on the adequate functioning of both male and female reproductive systems. The male testes must produce an adequate number of strong and active sperm cells. In women, sufficient egg production is the primary factor in fertility. Then, the vagina must be hospitable to the sperm, the cervical mucus must be thin enough for the sperm cells to enter, and the fallopian tube and uterus must contract, enabling the sperm to move towards the ovum.

Pelvic inflammatory disease, which can cause blockage of the fallopian tube, is a common cause of subfertility in women. Male subfertility is often due to a failure of the testes to manufacture sperm. The sperm may also lack normal shape and movement; this, too, can cause fertility problems.

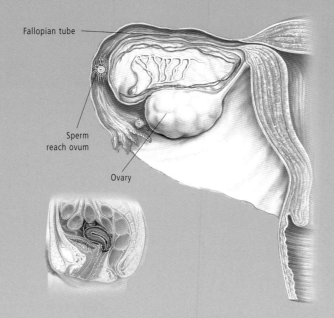

Fallopian tube

Sperm reach ovum

Ovary

Female fertility

Egg production and a favorable uterine environment are the key factors in female fertility. Once an egg has been produced successfully in the ovaries it moves down the fallopian tubes where it may be fertilized.

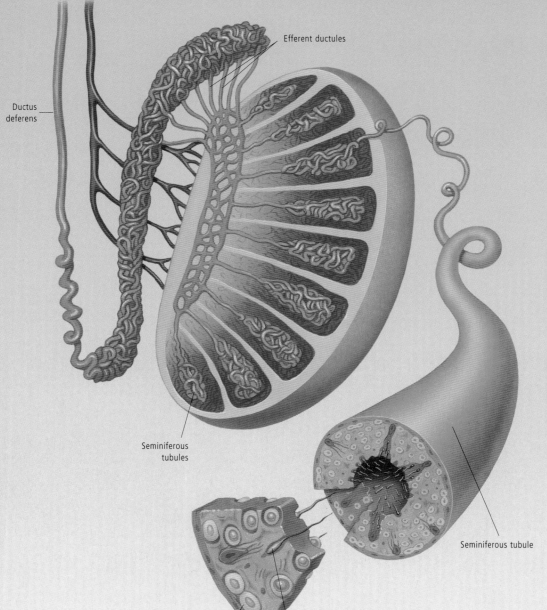

Efferent ductules

Ductus deferens

Seminiferous tubules

Spermatocyte

Spermatozoa

Seminiferous tubule

Male fertility

Male fertility is dependent on the production of an adequate number of vigorous sperm cells. Sperm are produced in the seminiferous tubules of the testes.

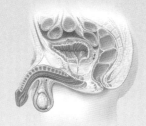

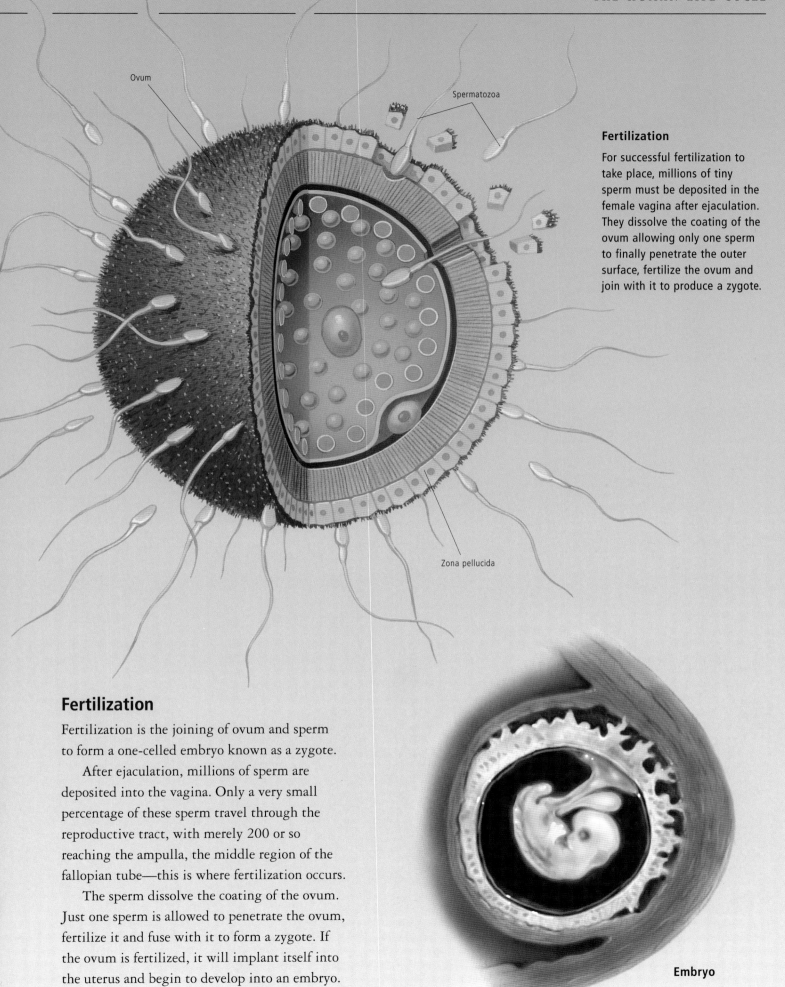

Ovum

Spermatozoa

Zona pellucida

Fertilization

For successful fertilization to take place, millions of tiny sperm must be deposited in the female vagina after ejaculation. They dissolve the coating of the ovum allowing only one sperm to finally penetrate the outer surface, fertilize the ovum and join with it to produce a zygote.

Fertilization

Fertilization is the joining of ovum and sperm to form a one-celled embryo known as a zygote.

After ejaculation, millions of sperm are deposited into the vagina. Only a very small percentage of these sperm travel through the reproductive tract, with merely 200 or so reaching the ampulla, the middle region of the fallopian tube—this is where fertilization occurs.

The sperm dissolve the coating of the ovum. Just one sperm is allowed to penetrate the ovum, fertilize it and fuse with it to form a zygote. If the ovum is fertilized, it will implant itself into the uterus and begin to develop into an embryo.

Embryo

Fetal Development

FETAL DEVELOPMENT: THE BRAIN

In the third week of an embryo's life, the neural tube has developed. This is a closed tube that will eventually form the brain and spinal cord.

During the 9-month gestation period, the front part of this tube enlarges: this becomes the proencephalon, the mesencephalon and the rhombencephalon, which eventually develop into the mature brain. The remainder of the neural tube becomes the spinal cord.

The relationship of these sections of the brain to each other will alter over the 9 months, as the brain flexes and folds up on itself.

DID YOU KNOW?

The fontanelle (or "soft spot") on a newborn's skull is a gap between the bones, composed of fibrous connective tissue. This gap closes naturally as the bones ossify.

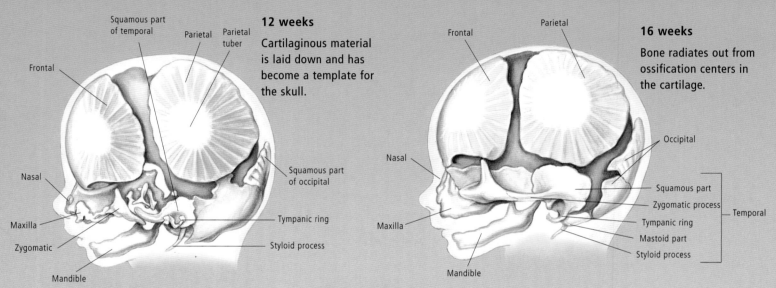

12 weeks
Cartilaginous material is laid down and has become a template for the skull.

16 weeks
Bone radiates out from ossification centers in the cartilage.

Fetal skull development

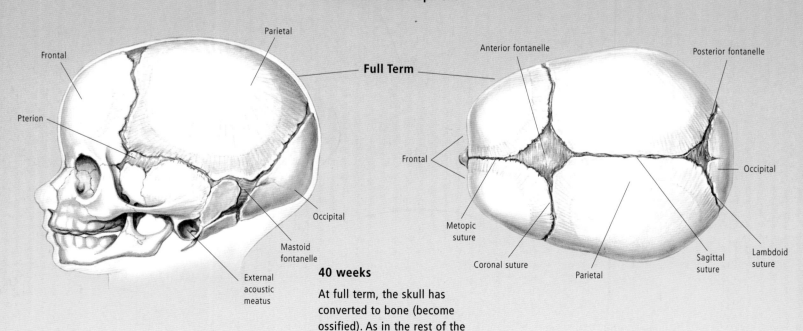

Full Term

40 weeks
At full term, the skull has converted to bone (become ossified). As in the rest of the skeleton, bone in the skull gradually spreads out to replace the cartilaginous material.

8 weeks

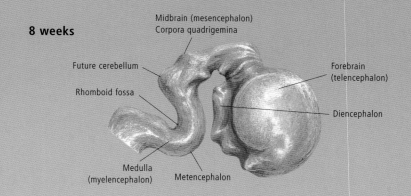

Midbrain (mesencephalon)
Corpora quadrigemina
Future cerebellum
Rhomboid fossa
Forebrain (telencephalon)
Diencephalon
Medulla (myelencephalon)
Metencephalon

11 weeks

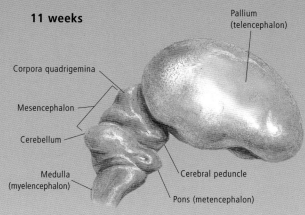

Pallium (telencephalon)
Corpora quadrigemina
Mesencephalon
Cerebellum
Medulla (myelencephalon)
Cerebral peduncle
Pons (metencephalon)

21 weeks

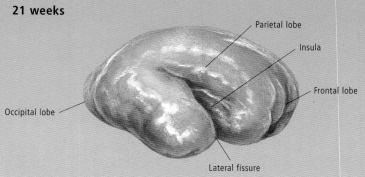

Parietal lobe
Insula
Frontal lobe
Occipital lobe
Lateral fissure

26 weeks

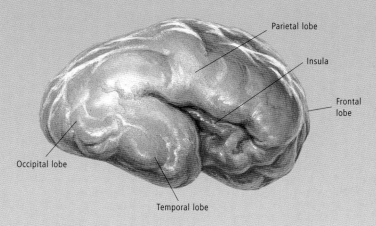

Parietal lobe
Insula
Frontal lobe
Occipital lobe
Temporal lobe

30 weeks

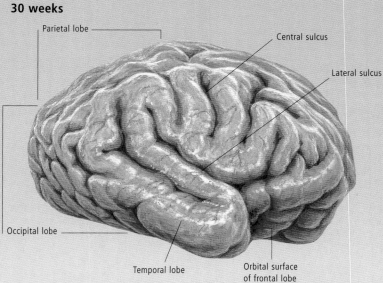

Parietal lobe
Central sulcus
Lateral sulcus
Occipital lobe
Temporal lobe
Orbital surface of frontal lobe

40 weeks

At birth, all the surface features of the adult brain are present.

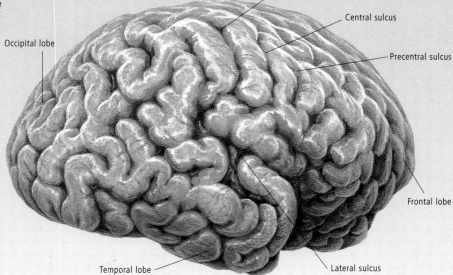

Occipital lobe
Postcentral sulcus
Central sulcus
Precentral sulcus
Frontal lobe
Temporal lobe
Lateral sulcus

Fetal brain development

Over a period of 9 months, the primitive neural tube forms into the prosencephalon, the mesencephalon and the rhombencephalon, which in turn develop into the various sections of the mature brain. By 4 weeks the prosencephalon has become the telencephalon and the diencephalon, and the rhombencephalon has become the myelencephalon and the metencephalon. The telencephalon develops into the cerebral hemispheres. At full term, all the surface features of the adult brain are already present.

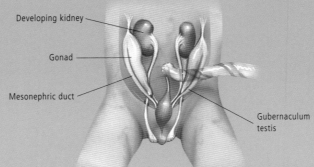

7 weeks

Developing kidney

Gonad

Mesonephric duct

Gubernaculum testis

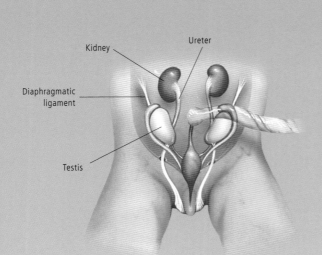

Kidney

Ureter

Diaphragmatic ligament

Testis

16 weeks

Descent of the testes

The embryonic male sex glands (testes) are formed from a piece of tissue in the abdomen, close to the kidneys. At about 30 weeks the testes are fully developed and begin to move down the inguinal canal. At full term, the testes have reached the scrotum.

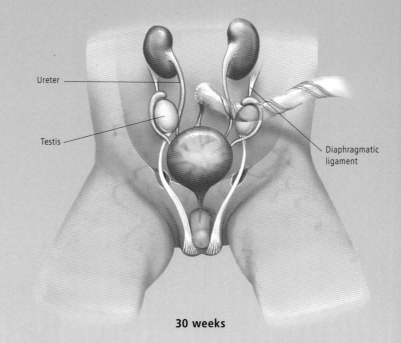

Ureter

Testis

Diaphragmatic ligament

30 weeks

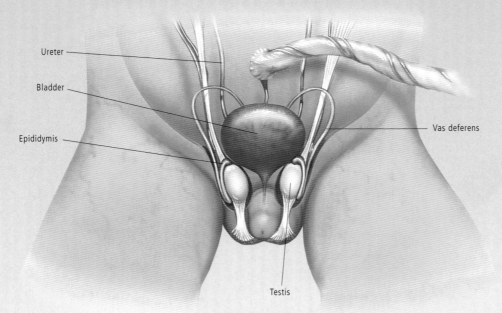

Ureter

Bladder

Epididymis

Vas deferens

Testis

Fully developed

DID YOU KNOW?

The ovum contains 23 single X chromosomes, and the sperm contains either 23 X or 23 Y chromosomes. If the ovum is fertilized by Y chromosomes, the baby will be a boy; if fertilized by X chromosomes, it will be a girl.

SEX DIFFERENTIATION IN THE FETUS

Although the genitals begin to develop in the fetus in the second month, it is not until week 12 that the different male and female characteristics can be seen.

Up until week 12, the genital organs simply consist of a genital tubercle, a urogenital membrane, two urogenital folds and a swelling on each side of the fold.

By the twelfth week, the tubercle and the urogenital folds become the penis, and the swellings come together to form the scrotum. If the fetus is a female, however, the tubercle becomes the clitoris and the folds and swellings become the lips of the vulva.

Over time, both the male testes and the female ovaries make their way further down the body. The testes are in place in the scrotum by the end of the eighth month. A pair of tubules form—these join with the testes and open into the urethra, and pouches in the ducts become the seminal vesicles.

A pair of ducts also develops in the female; one end comes to lie next to the ovaries, and the other end fuses into a tube that becomes the uterus and vagina.

Fetal sex differentiation

Up to 12 weeks, there is no difference between the male and female external genital organs, which consist of a genital tubercle, a uro-genital membrane, a pair of urogenital folds and a genital swelling. After 12 weeks, these differentiate into the penis and scrotum in the male, and the clitoris and vulva in the female.

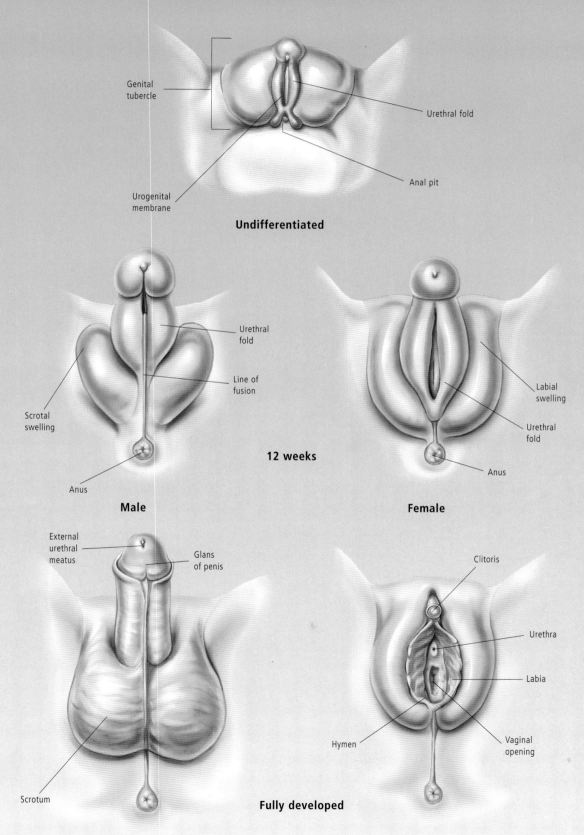

Undifferentiated

12 weeks

Male

Female

Fully developed

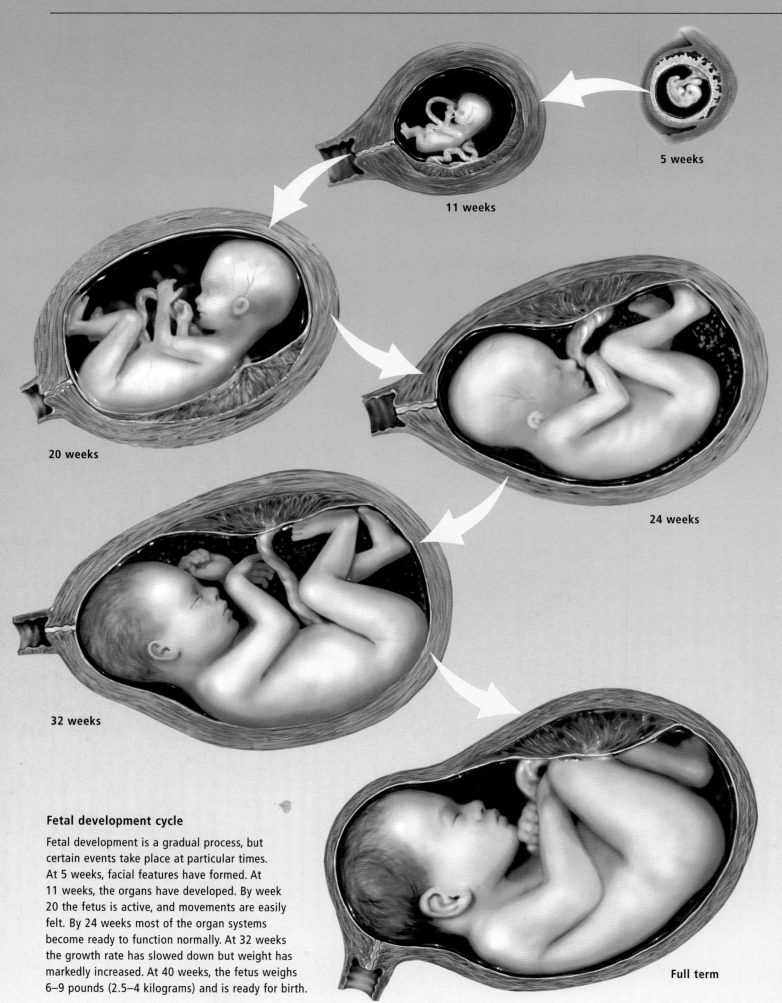

5 weeks

11 weeks

20 weeks

24 weeks

32 weeks

Full term

Fetal development cycle

Fetal development is a gradual process, but certain events take place at particular times. At 5 weeks, facial features have formed. At 11 weeks, the organs have developed. By week 20 the fetus is active, and movements are easily felt. By 24 weeks most of the organ systems become ready to function normally. At 32 weeks the growth rate has slowed down but weight has markedly increased. At 40 weeks, the fetus weighs 6–9 pounds (2.5–4 kilograms) and is ready for birth.

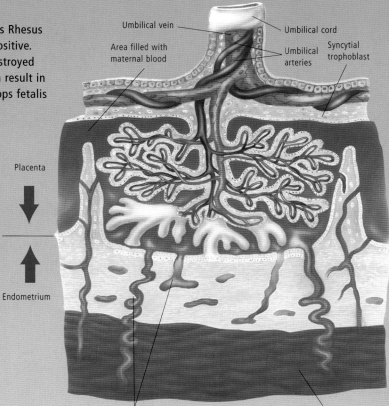

Umbilical vein
Area filled with maternal blood
Umbilical cord
Umbilical arteries
Syncytial trophoblast
Placenta
Endometrium
Maternal blood vessels
Myometrium

Rhesus Factor

Difficulties can arise if the mother is Rhesus (Rh) negative and the fetus is Rh-positive. The fetal red blood cells may be destroyed by the mother's antibodies; this can result in jaundice, neonatal anemia, or hydrops fetalis (swelling due to excessive fluid).

First pregnancy

In the first pregnancy, problems do not usually occur because although a small number of fetal red blood cells enter the maternal circulation, these are rapidly destroyed before they can provoke an antibody response.

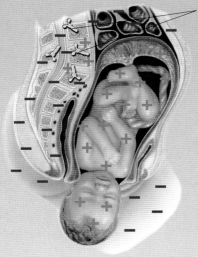

Placenta keeps maternal and fetal blood separate

Fetus positive

Mother Rh-negative

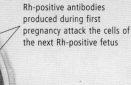

Antibodies developing

First pregnancy: birth

During childbirth, a much larger amount of baby's blood can leak into the mother's bloodstream, and this is more likely to provoke antibody production. The Rh-negative mother develops antibodies that destroy the Rh-positive blood cells. Rh-positive antibodies produced in the first pregnancy attack the cells of the second Rh-positive fetus.

Placenta

The placenta joins the mother and offspring via the umbilical cord. It holds the baby in place, distributes nutrients and oxygen to the baby, and removes fetal waste. The placenta also produces hormones.

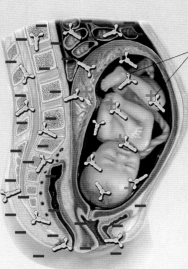

Rh-positive antibodies produced during first pregnancy attack the cells of the next Rh-positive fetus

Further pregnancies

The mother needs to receive an Anti-D globulin injection following the first pregnancy to stop production of antibodies. If she does not, her antibodies may attack the blood cells of subsequent Rh-positive babies.

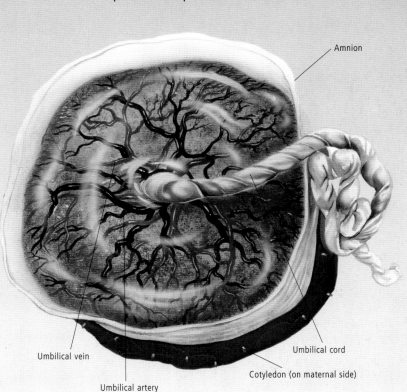

Amnion

Umbilical cord

Cotyledon (on maternal side)

Umbilical artery

Umbilical vein

Childbirth

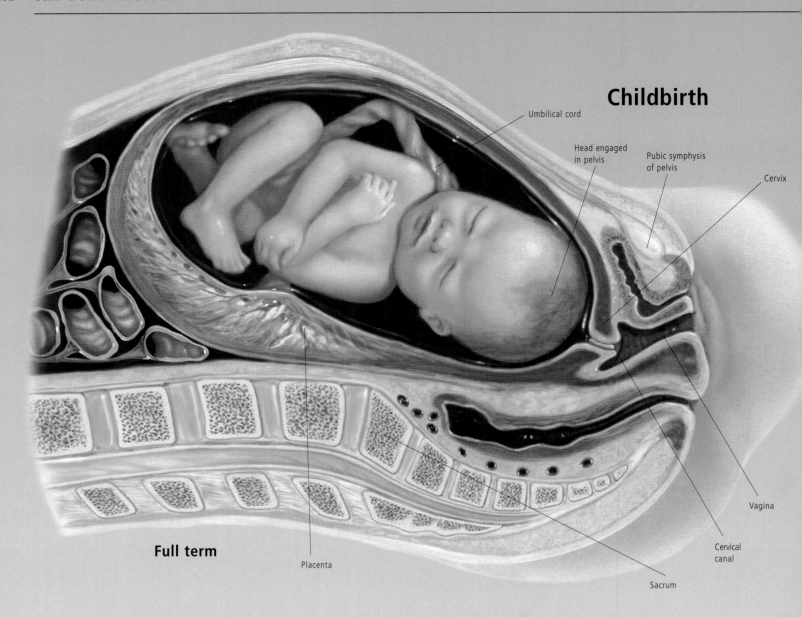

Umbilical cord

Head engaged in pelvis

Pubic symphysis of pelvis

Cervix

Full term

Placenta

Vagina

Cervical canal

Sacrum

Breech birth

In a breech birth (6 percent of deliveries), the baby's head faces the mother's pubis, and the legs and buttocks enter the birth canal first. The midwife or obstetrician sometimes turns the baby in the uterus—the baby is then born naturally. If turning the baby fails, or is not attempted, elective cesarean section will generally be required.

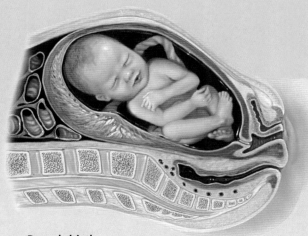

Twins

Fraternal twins develop with separate placentas, and may be of different sexes. Identical twins are produced often after the fertilized egg has implanted itself in the lining of the uterus, where it splits in two. They share a placenta, and are of the same sex.

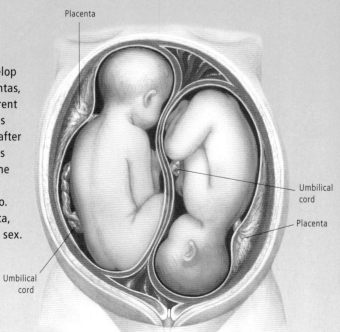

Placenta

Umbilical cord

Placenta

Umbilical cord

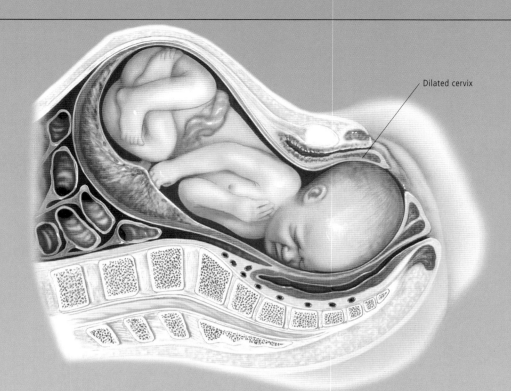

Dilated cervix

First stage: Dilation

The cervix must fully dilate to enable the baby to move into the birth canal. Hormonal activity helps the uterus to contract, as it works to shorten and soften the cervix by pressing the baby's head against it. Once the cervix is completely dilated, the baby can move into the birth canal.

Second stage: Crowning

The mother bears down and pushes the baby into the birth canal. After several pushes, the baby's head presents at the opening of the vagina: this is known as the "crowning".

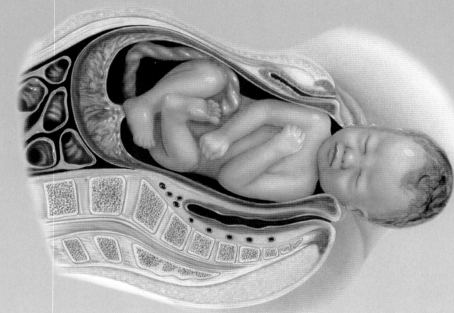

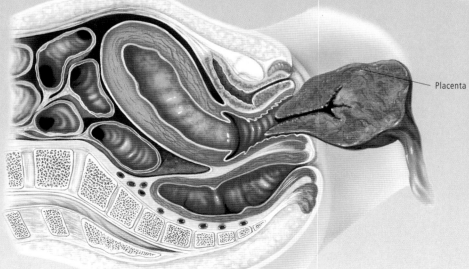

Placenta

Third stage: Expulsion of placenta

The baby is born, but the uterus continues to contract; a hormonal rush signals the placenta to peel away from the uterus. Once the placenta has separated, it will slide down the birth canal.

DNA

Deoxyribonucleic acid (DNA) molecules, found in the chromosomes in the nucleus of every cell, carry the genetic information that determines inherited traits. This information controls the formation of proteins used by the body for growth and chemical processes. Unraveling a chromosome shows that DNA comprises a spiral ladder consisting of two chains of phosphate and sugar units attached to nitrogenous bases. The chains are joined together at the bases like the rungs of a ladder. There are four bases in DNA: adenine, thymine, cytosine and guanine. These bases bind together in limited combinations (adenine always binds with thymine; cytosine always binds with guanine). Proteins are made up of chains of amino acids. DNA passes on its information to protein factories (ribosomes) in the cytoplasm of the cell by creating a messenger acid—messenger ribonucleic acid (mRNA).

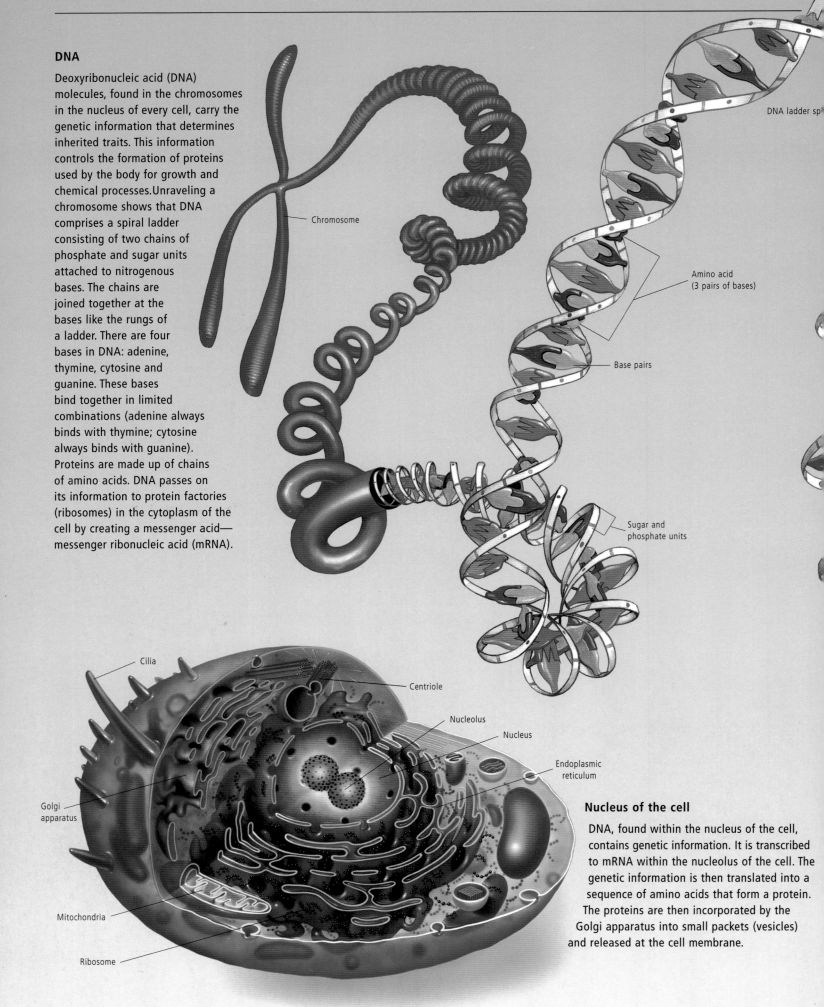

Chromosome

DNA ladder sp

Amino acid
(3 pairs of bases)

Base pairs

Sugar and
phosphate units

Cilia

Centriole

Nucleolus

Nucleus

Endoplasmic
reticulum

Golgi
apparatus

Mitochondria

Ribosome

Nucleus of the cell

DNA, found within the nucleus of the cell, contains genetic information. It is transcribed to mRNA within the nucleolus of the cell. The genetic information is then translated into a sequence of amino acids that form a protein. The proteins are then incorporated by the Golgi apparatus into small packets (vesicles) and released at the cell membrane.

Cytosine   Adenine   Thymine

Guanine   Uracil

Nucleotide bases

Genetic Information

Each nucleotide base interlocks with a specific partner to form a base pair. Three base pairs form a codon and code for one amino acid. The order in which the bases are carried determines the information contained in that strand. Genetic information is contained in the many combinations of bases that exist along the length of the DNA molecule. A gene is a particular sequence of bases which codes for a specific protein. Proteins catalyze chemical reactions, build cells and tissues, and pass on characteristics to an individual.

Genetic Instructions

Most characteristics are the result of a combination of two sets of genetic instructions contained in one or more pairs of genes, but some features, such as eye color, are determined by a single gene. This dominant gene overrides the instructions of the other recessive gene. A recessive trait can only come through when two recessive genes for that trait are inherited. The gene for brown eyes is dominant over the recessive gene for blue eyes. Two parents with brown eyes can only have a child with blue eyes if the child inherits a recessive blue gene from each parent (a). If one parent with brown eyes has two dominant brown-eye genes, all children will inherit at least one dominant gene and will all have brown eyes (b). If both parents have blue eyes, neither will have the dominant gene and all children will have blue eyes.

Uracil

mRNA

The strands of DNA rejoin

a

Mother   Father

Children

b

Mother   Father

Children

4 months

At 4 months, most babies are able to sit up for a brief period without support, and can reach out and grasp objects.

8 months

Babies generally begin to crawl between 7 and 10 months.

Infancy

Infants develop at their own rate, but there is a specific order of developmental changes. For example, babies must learn the muscular skill of head control before they learn to sit up.

Infancy

DEVELOPMENTAL STAGES

In the first 12 months, more growth and development takes place than at any other time of life. During this important first year, infants gain in height and weight, begin to walk and talk, and start teething. An infant's brain virtually doubles in weight in the first year as the number of brain cells increases.

Infants progress at their own pace, but there is a specific order of developmental changes because of the way the body and nervous system mature. Babies must learn the muscular skill of head control, for example, before they can sit up.

Some babies develop physically more rapidly than others; other babies develop verbal or social skills sooner than those of the same age. Crying and smiling are the first means of communication—a baby will smile during the first 6 weeks. A baby begins to mimic sounds from birth, but will not be able to utter actual words until 12–18 months.

When infants exercise their reflexes by sucking, grasping, kicking and throwing, they are actually beginning to explore and make sense of their surroundings. In the first 3 months, the senses develop very quickly. At 3 months the infant can distinguish form and color.

Generally, the teeth start to appear at about 6–9 months. By the end of the first year, up to 12 of the 20 primary or baby teeth are in place. Teething can begin as early as 3 months and end as late as 3 years.

A baby's sleep is more restless than an adult's because a baby's brain is more alert during sleep. Babies are not able to fall into a deep sleep until the end of the third month; by 6 months they are able to sleep for longer periods.

12 months

At 12 months, most infants are able to stand up alone; at this age, many infants are also beginning to take their first steps.

Premature babies are those born before Week 37. A baby born 4 weeks early will be developmentally 4 weeks behind a full-term baby born at the same time.

It is important that babies have regular checkups to make sure they are growing and developing normally.

Childhood

Childhood is the period between infancy and adolescence. It is a time of rapid physical and intellectual growth during which education and environment build a foundation for later years.

Each child is individual and will progress differently, depending on genetic and social influences; however, the changes are ordered and specific.

By age 2, a child will be able to walk alone, greet people, drink from a cup and comprehend simple commands and questions. By age 3 or 4, most children will be capable of walking, jumping, running, climbing stairs and handling objects. They can form short sentences and will have a vocabulary of several hundred words. In the pre-school years, the child develops cognitive skills such as thinking, recognizing and remembering.

Most children grow in irregular spurts. After the first year, a child's growth slows considerably, and by 2 years growth continues fairly steadily, with height increasing about $2^1/_2$ inches (6 centimeters) per year until the teenage years. Normal growth depends on adequate nutrition, exercise and rest. Most children need 10 to 12 hours of sleep per night.

By age 6, all the primary teeth would have appeared, the secondary teeth would have begun to emerge, gradually replacing the primary teeth.

The long bones, such as those in the arm and leg, begin as cartilage "models" which are then converted to bone.

As the brain grows, the flat bones of the skull enlarge by expanding at the margins. The brain, along with the rest of the head, develops earlier than the rest of the body. At age 6 the brain has reached about 90 percent of its adult weight.

Developmental milestones are guides only, but parents who are worried that their child may not be progressing normally should seek medical advice. However, most children who are small or developmentally delayed, are normal and healthy.

Developmental Stages

Children develop at varying rates. Some children are very advanced intellectually, while others will develop physically or socially ahead of their peers.

5-year-old boy

5-year-old girl

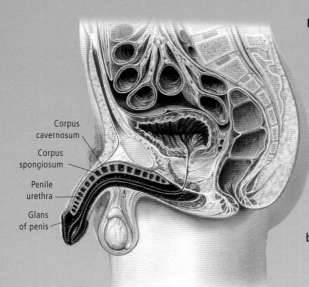

Corpus
cavernosum

Corpus
spongiosum

Penile
urethra

Glans
of penis

Penis

The penis is composed of three cylinders. Two cylinders (the corpus cavernosa) are made of sponge-like vascular tissue which allow erection. The third cylinder (the corpus spongiosum) contains the urethra—part of the urinary system. The penis is attached to the pelvic bone by connective tissue, and usually hangs flaccid unless sexually stimulated. During puberty, there is rapid growth in both the length and girth of the penis.

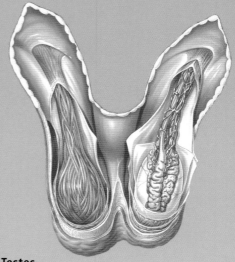

Testes

Situated outside the body in the scrotum, the testes lie directly behind the penis. They are the major organs of reproduction in the male, and produce sperm and the male sex hormones, including testosterone.

Male Puberty

During puberty, the secondary sexual characteristics begin to develop, and the reproductive organs become fully functional. Boys usually reach puberty at age 13–14. Over the next several years, there is an increase in body size, change in body shape, and development of the reproductive organs.

The male secondary sexual characteristics include rapid growth in the size of the testes and penis, an increase in the size of the larynx, and the appearance of facial, body, underarm and pubic hair. Height also increases, and about a year or so after the penis begins to enlarge the first ejaculation occurs. These changes are triggered by the hormones released by the pituitary gland, which enable the testes to produce sperm and testosterone.

Pituitary

Larynx

Trachea

Esophagus

Testes

Voice box

During puberty, the size of the voice box (larynx) increases, thus deepening the voice and giving the "Adam's Apple" appearance to the front of the neck.

Testosterone

All of the changes in puberty are governed by the male sex hormone, testosterone. The hormone is produced by the testes, under the control of the pituitary gland. Testosterone is also important in the body's metabolism and muscle growth.

Testosterone is secreted by the testes; the production of testosterone is triggered by the follicle stimulating hormone (FSH) and luteinizing hormone, both of which are secreted by the anterior lobe of the pituitary gland.

There is increased glandular activity in both sexes, with the apocrine glands becoming active. The sebaceous glands of both sexes also become more active, and this can cause acne.

Emotional and behavioral fluctuations accompany the physical changes of puberty. Sexual feelings are aroused, and many adolescents start to become sexually active. Dating usually begins during puberty.

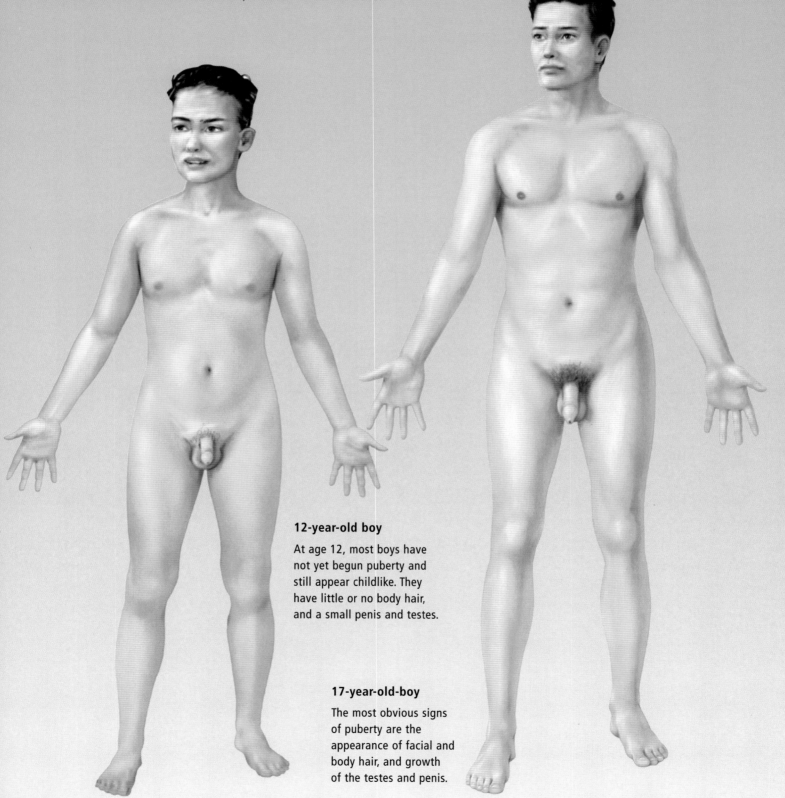

12-year-old boy

At age 12, most boys have not yet begun puberty and still appear childlike. They have little or no body hair, and a small penis and testes.

17-year-old-boy

The most obvious signs of puberty are the appearance of facial and body hair, and growth of the testes and penis.

Female Puberty

Puberty in girls begins around age 11 and continues until about age 16. Girls go through greater physical changes than boys, but take less time to reach maturity. These physical changes include development of the breasts, widening of the hips, rapid growth of the uterus, and the appearance of hair on the underarms and around the vulva.

As well as the outward physical changes, there are also hormonal changes caused by the actions of estrogen and progesterone, released by the ovaries, under the control of the pituitary gland.

The uterus reaches its adult shape and the first menstrual period occurs. The onset of menstruation is called menarche. Menstruation, a monthly discharge of blood and uterine tissue, begins approximately 2 years after the onset of puberty. During the first few years, periods are usually irregular.

Acne and eating disorders are more prevalent at this time than at any other time in life.

Inherited tendencies, nutrition and the environment all influence the times and rates of growth and sexual maturity.

Ovulation

Ovulation takes place midway through each menstrual cycle. A Graafian follicle ruptures to release its ovum, which enters the fallopian tube and travels towards the uterus. The ovum is either fertilized here, or degenerates within the next few days.

After ovulation, the secretory cells remaining in the ovary become a corpus luteum, producing progesterone and small amounts of estrogen until day 26 of a non-pregnant cycle. The corpus luteum then deteriorates, with a rapid reduction in estrogen and progesterone secretion, resulting in menstruation.

Breast

The breasts, or mammary glands, are composed mainly of fat cells interspersed with sac-like structures called lobules. When stimulated by certain hormones such as prolactin, the lobules can produce milk in females. The lobules

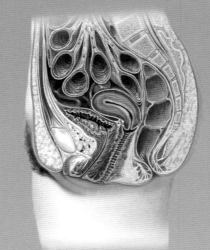

Female reproductive system

The vagina is a long fibro-muscular tube that extends from the cervix to the vulva. The uterus is located behind the bladder and in front of the rectum. The upper two-thirds of the uterus is called the body, and the lower third is known as the cervix.

Ovulation

When the ovary releases an ovum, usually on Day 14 of the menstrual cycle, this is known as ovulation.

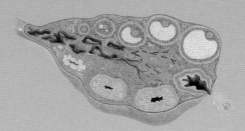

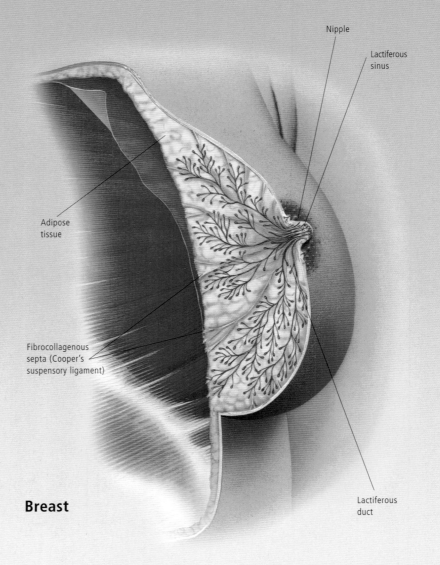

Nipple

Lactiferous sinus

Adipose tissue

Fibrocollagenous septa (Cooper's suspensory ligament)

Lactiferous duct

Breast

empty into a network of ducts, which form channels that carry milk to the nipple.

During puberty, the hormone estrogen causes the breasts to grow in size, with size and shape generally determined by inherited factors. Towards the end of each menstrual cycle the breasts may temporarily swell and become tender. The breasts enlarge during pregnancy due to an increase in estrogen.

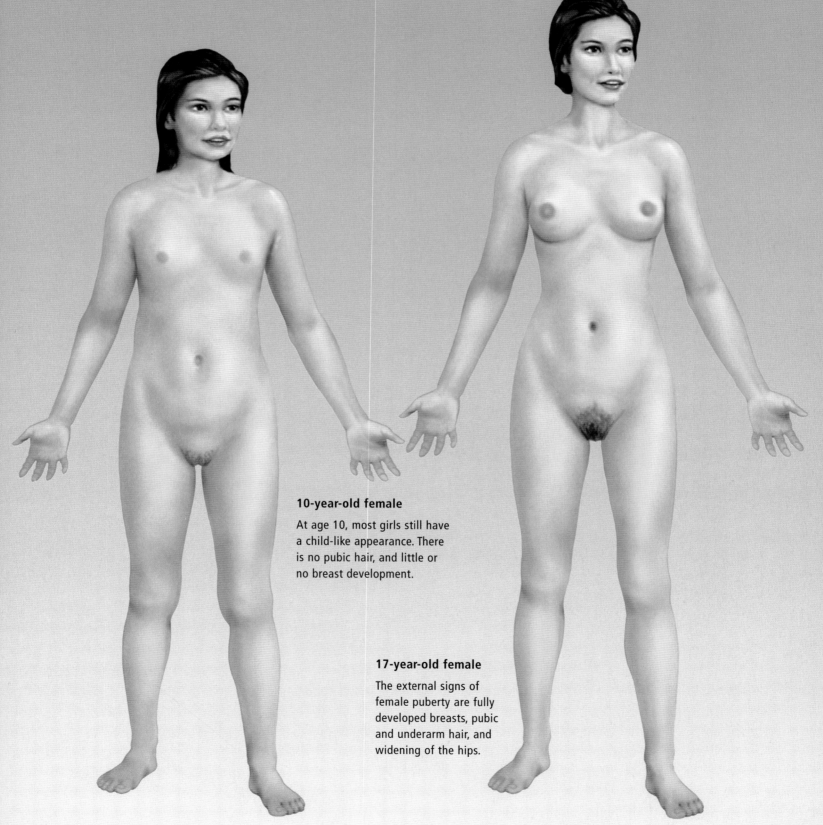

10-year-old female

At age 10, most girls still have a child-like appearance. There is no pubic hair, and little or no breast development.

17-year-old female

The external signs of female puberty are fully developed breasts, pubic and underarm hair, and widening of the hips.

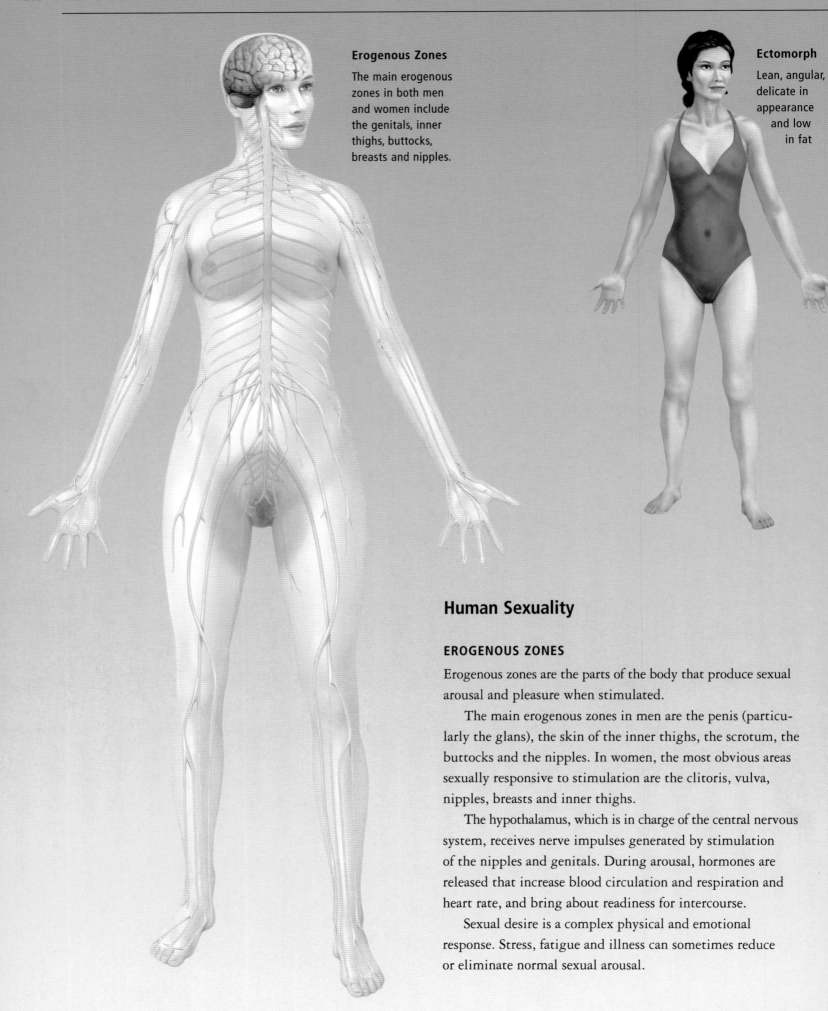

Erogenous Zones

The main erogenous zones in both men and women include the genitals, inner thighs, buttocks, breasts and nipples.

Ectomorph

Lean, angular, delicate in appearance and low in fat

Human Sexuality

EROGENOUS ZONES

Erogenous zones are the parts of the body that produce sexual arousal and pleasure when stimulated.

The main erogenous zones in men are the penis (particularly the glans), the skin of the inner thighs, the scrotum, the buttocks and the nipples. In women, the most obvious areas sexually responsive to stimulation are the clitoris, vulva, nipples, breasts and inner thighs.

The hypothalamus, which is in charge of the central nervous system, receives nerve impulses generated by stimulation of the nipples and genitals. During arousal, hormones are released that increase blood circulation and respiration and heart rate, and bring about readiness for intercourse.

Sexual desire is a complex physical and emotional response. Stress, fatigue and illness can sometimes reduce or eliminate normal sexual arousal.

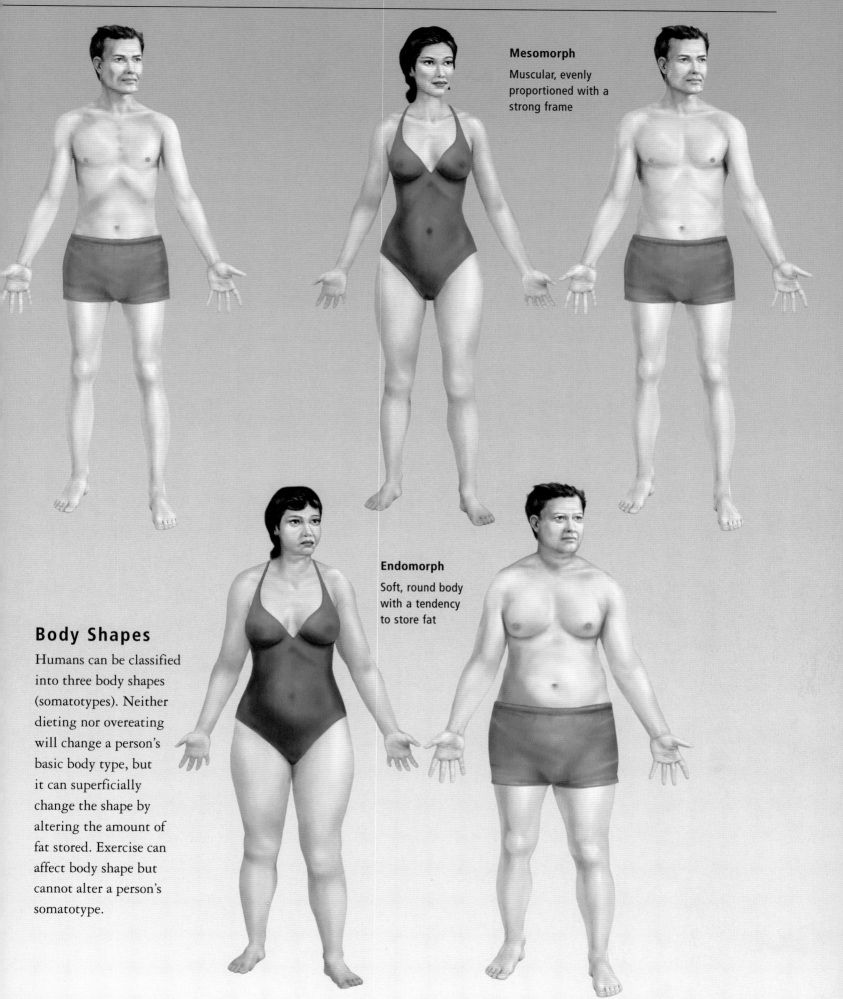

Mesomorph
Muscular, evenly
proportioned with a
strong frame

Endomorph
Soft, round body
with a tendency
to store fat

Body Shapes

Humans can be classified
into three body shapes
(somatotypes). Neither
dieting nor overeating
will change a person's
basic body type, but
it can superficially
change the shape by
altering the amount of
fat stored. Exercise can
affect body shape but
cannot alter a person's
somatotype.

Metabolism

Metabolism is the general term for the chemical processes that run the body. Thyroid hormones regulate the metabolic rate, so an over-active or an under-active thyroid can upset the body's metabolism. The pituitary gland governs the release of thyroid hormones.

There are two basic phases or processes involved in metabolism—one is constructive and uses energy (anabolism), the other breaks down compounds and creates energy (catabolism).

During anabolism, simple molecules are converted into more complex molecules and substances. Anabolic reactions require energy and occur during the growth, repair and maintenance of body cells and systems.

During catabolism, food is converted into simpler compounds, producing energy which is then stored ready for use by the body.

· Enzymes and nutrients are essential for the chemical reactions that take place in our bodies. Enzymes are produced by the body while nutrients are derived from food.

Regulating body temperature

Both the dermal and epidermal layers of the skin are involved in regulating body temperature. When the body is hot, arteries dilate and blood flow to the skin increases, accelerating heat loss. The sweat glands are activated and release fluid, which evaporates and reduces temperature. When the body is cold, the arteries and pores contract and the hairs of the skin become erect, providing an insulating layer, trapping body heat close to the skin's surface.

Hypothalamus

Signals travelling via the nerves direct blood flow to organs or the skin, depending on the temperature of the body

Nerve receptors in the skin relay messages to the hypothalamus on body temperature

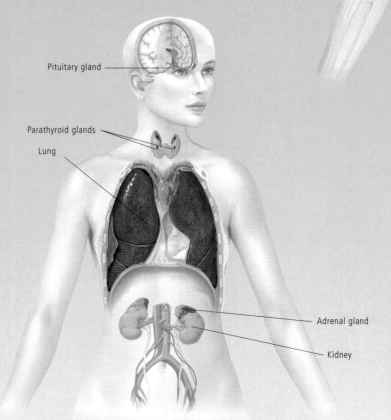

Pituitary gland

Parathyroid glands

Lung

Adrenal gland

Kidney

DID YOU KNOW?

An adult's normal body functions require around 1,500–2,000 kilocalories (6,000–8,000 kilojoules) each day. Generally the metabolic rate of men is higher than that of women.

Electrolytes

The concentration of electrolytes in the blood is regulated by various organs. The adrenal glands regulate sodium and potassium; the pituitary gland releases hormones that contribute to the regulation of electrolyte balance; the parathyroid glands regulate calcium and phosphate; the kidneys and lungs regulate bicarbonate; and the kidneys regulate chloride.

Temperature control

There is a mechanism in the body for maintaining a stable temperature. This mechanism is controlled by the hypothalamus in the brain. A change in outside temperature is relayed to the hypothalamus by nerve endings in the surface of the skin. If the hypothalamus receives a signal that the body is cold, it increases heat production in the body by increasing the rate of metabolism. If the body is hot, the hypothalamus causes arteries to dilate, sending blood to the skin where heat can be lost.

Skin reaction to heat

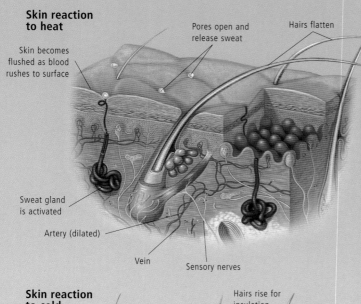

Skin becomes flushed as blood rushes to surface

Pores open and release sweat

Hairs flatten

Sweat gland is activated

Artery (dilated)

Vein

Sensory nerves

Skin reaction to cold

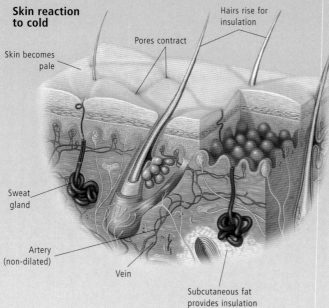

Hairs rise for insulation

Pores contract

Skin becomes pale

Sweat gland

Artery (non-dilated)

Vein

Subcutaneous fat provides insulation

Carbohydrates into glucose

One of the most important processes in the body is the breakdown of carbohydrates into glucose, and the conversion of glucose into energy. Chemicals in the stomach, intestines and pancreas break down carbohydrates and other nutrients. The liver metabolizes food and stores glucose to provide energy for muscles and cells.

Parathyroid

Bones store calcium

Kidneys excrete calcium

Negative feedback

The body uses the mechanism of the negative feedback loop to regulate hormone levels. Too much or too little of a particular hormone triggers a response to normalize hormone levels. Calcium levels in the blood are controlled by parathyroid hormones that instruct the bones to store calcium or the kidneys to excrete it, depending on circulating levels. If levels go down too far, the parathyroid registers the deficit and releases parathyroid hormone, which tells the bones to release calcium, increasing blood calcium levels.

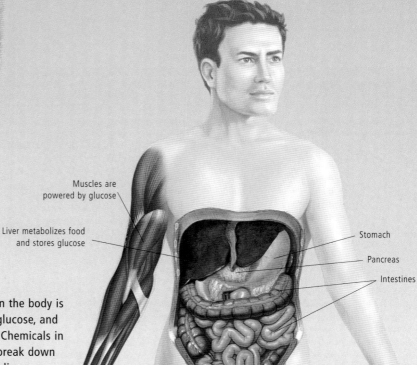

Muscles are powered by glucose

Liver metabolizes food and stores glucose

Stomach

Pancreas

Intestines

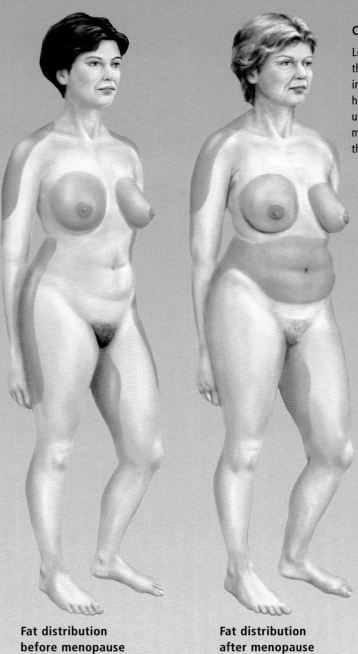

**Fat distribution
before menopause**

**Fat distribution
after menopause**

Changes in fat distribution

Loss of estrogen and progesterone during the menopausal years affects fat distribution in women. Before menopause, most women have some fat deposits around the hips, thighs, upper arms and breasts. After menopause, most of the fat is concentrated around the abdomen, waist and breasts.

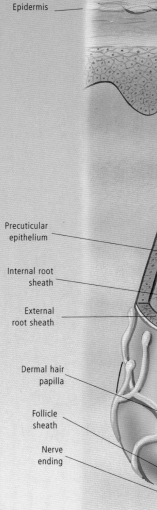

Epidermis

Precuticular
epithelium

Internal root
sheath

External
root sheath

Dermal hair
papilla

Follicle
sheath

Nerve
ending

Menopause

Menopause, or "climacteric", is the time of a woman's life during which ovarian function and hormonal production decline. The whole process, which can last for several years (usually starting in the 40s), involves three phases. The premenopause can refer to most of a woman's reproductive years, but more commonly includes the years when the menstrual periods become heavy and irregular. During the second phase (the perimenopause) a woman may experience symptoms such as hot flashes, depression, sleep problems, night sweats and vaginal dryness. Actual menopause is said to start after 12 months have passed without a menstrual period (approximately at age 50).

Hair follicle

The shaft of the hair projects from the surface of the skin. The root is embedded in the skin; it ends in a bulb, which is lodged in a pit known as the follicle.

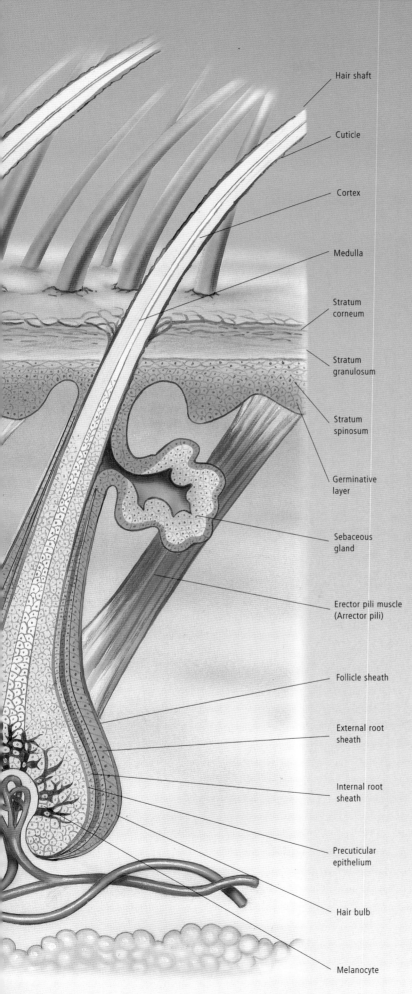

Hair shaft

Cuticle

Cortex

Medulla

Stratum
corneum

Stratum
granulosum

Stratum
spinosum

Germinative
layer

Sebaceous
gland

Erector pili muscle
(Arrector pili)

Follicle sheath

External root
sheath

Internal root
sheath

Precuticular
epithelium

Hair bulb

Melanocyte

Once a woman has gone through menopause, the ovaries will no longer produce estrogen and progesterone. Estrogen is needed to maintain healthy body tissues; long-term estrogen deficiency can lead to osteoporosis, changes to the urogenital tissue, poor bladder control, dry skin, weight gain, loss of muscle strength, and increased risk of stroke and heart disease.

Some women may benefit from hormone replacement therapy; other women may find alternative therapies more suitable. Good nutrition, exercise, emotional support and recognition of changes will ease a woman through this natural life transition.

MALE MENOPAUSE

Some medical authorities believe that male menopause is a physical condition resulting from fluctuating hormone levels, especially testosterone. By age 50, it is common for the prostate to show signs of enlargement.

Male pattern baldness

The most common cause of hair loss (alopecia) is a fluctuation in male hormones. The most common form of baldness is male pattern baldness (androgenetic alopecia), an inherited condition associated with sexual development. Baldness begins as hair loss exceeds replacement rates. In men, baldness is age related, affecting 30 percent of men in their thirties, 40 percent of men in their forties, and so on.

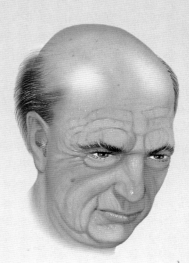

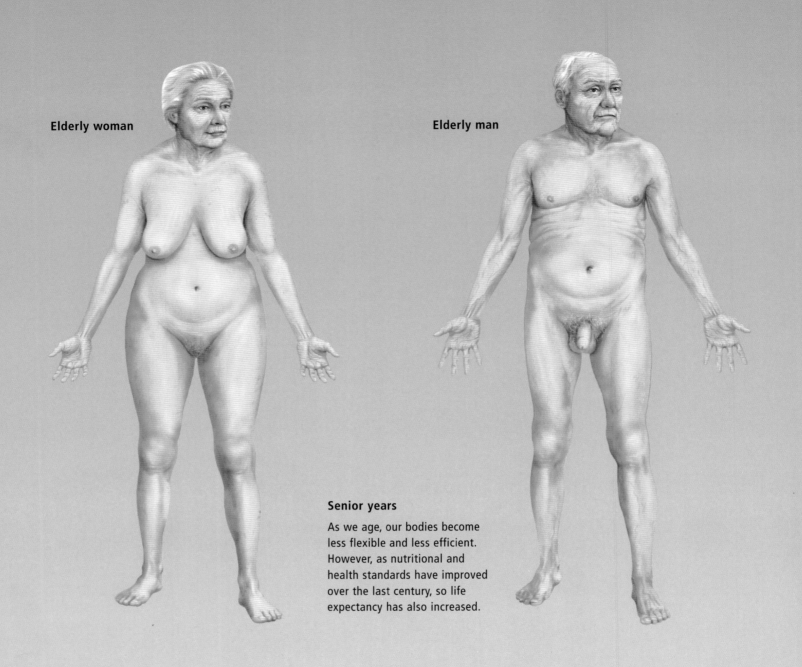

Elderly woman

Elderly man

Senior years

As we age, our bodies become less flexible and less efficient. However, as nutritional and health standards have improved over the last century, so life expectancy has also increased.

Senior Years

Although there is ample reason to suppose that an active lifestyle and balanced nutrition can maintain good health and fitness as we get older, it is also true that many diseases and disorders come with advancing age. The effects of age can be reduced and delayed, but they cannot be eliminated.

As people grow older, soft tissues become less flexible and the internal organs lose their efficiency. Visual impairment reduces the ability to drive, read and shop. Hearing high-pitched tones becomes more difficult, affecting how speech is heard and understood. Respiratory and circulatory ailments become more common. Blood flow to the liver, kidneys and brain can be reduced, affecting the elimination of waste products. Lung capacity decreases. Mental impairment is a common problem, affecting many aspects of an elderly person's life. Decreased balance and coordination, muscle weakness and arthritis may restrict movement.

Outward physical appearance also changes, skin loses its elasticity, becoming wrinkled; bone loses density, and spongy bones such as vertebrae can partially collapse, reducing height.

The immune system is less efficient, making older people more susceptible to infections and cancers. Other common problems that affect the elderly include fracture, heart attack, osteoporosis, Parkinson's disease, prostate problems and stroke.

INDEX

Introduction
References to figures and captions are in *italic*, references to
major discussions of a topic are in **bold**.